Gillies & Dodds
Bacteriology Illustrated

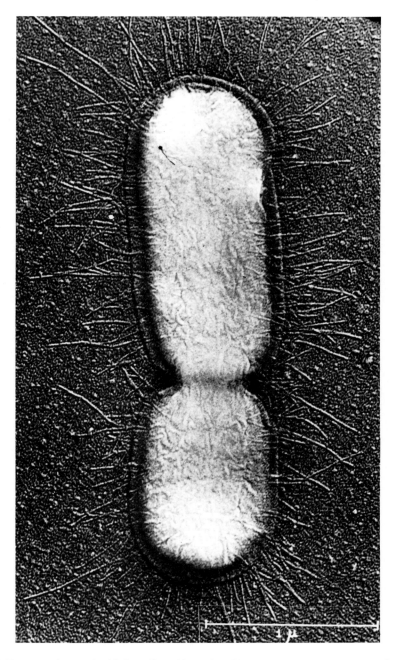

Frontispiece (1) Electron micrograph of *Salmonella typhi* × 16 000. The dividing bacillus possesses 12 flagella and approximately 100 fimbriae.

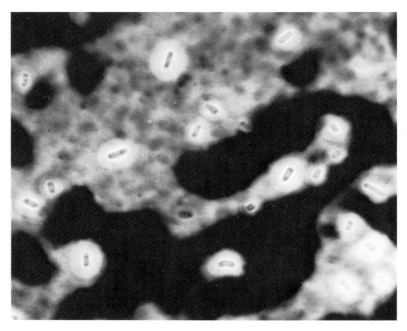

Frontispiece (2) *Klebsiella aerogenes* strain A3, capsular serotype 54 grown for 7 days on maltose peptone agar. Capsules and loose slime in wet film with India ink. × 2000.

Within each large capsule lies a central protoplast surrounded by a narrow, bright diffraction halo; the capsulate cells lie on a background of loose slime which is slightly darker than the capsule since it is partly and patchily infiltrated by the carbon particles of the ink. In contrast to the capsules the variety of shapes and arrangements of loose slime should be noted.

Gillies & Dodds Bacteriology Illustrated

Revised by

R. R. Gillies

MD, DPH, MRCPath MRCP(Ed)

Professor of Clinical Bacteriology, Queen's University of Belfast,
Belfast, UK

FIFTH EDITION

Churchill Livingstone

EDINBURGH LONDON MELBOURNE AND NEW YORK 1984

CHURCHILL LIVINGSTONE
Medical Division of Longman Group Limited

Distributed in the United States of America by Churchill
Livingstone Inc., 1560 Broadway, New York, N.Y.
10036, and by associated companies, branches and
representatives throughout the world.

First Edition 1965
Second Edition 1968
Third Edition 1973
Fourth Edition 1976
Fifth Edition 1984

ISBN 0 443 02809 5

British Library Cataloguing in Publication Data
Gillies, R. R.
    Bacteriology illustrated.—5th ed
    1. Bacteriology, Medical
    I. Title
    616′.014   QR46

Library of Congress Cataloging in Publication Data
Gillies, R. R. (Robert Reid)
    Bacteriology illustrated.

    Bibliography: p.
    Includes index.
    1. Bacteriology, Medical.   2. Bacteriology.
3. Protozoology.   4. Mycology.   I. Title. [DNLM:
1. Bacteriology.   QW 4 G481b]
QR46.G48 1983   616′.014   82–23595

Printed in Hong Kong
by C & C Joint Printing Co., (H.K.) Ltd.

Preface to the fifth edition

This volume, like previous editions, is the result of co-operation with many people; Mr J. P. Evans, Chief Technician in my Belfast Department, prepared material for several macroscopic and microscopic illustrations and these were photographed respectively by Mr Norman McMullan of the Medical Photography Department in the Royal Victoria Hospital, Belfast and by Mr J. Paul of the Medical Photography Unit in the University Medical School, Edinburgh. I am especially indebted to Mr Paul for his collaboration and expertise in the final stages of preparing material for publication.

The Ulster-Scottish connection was also invaluable in that Mr Brendan Ellis of the Royal Victoria Hospital prepared some fresh illustrations whilst Mr Ian Lennox of the Edinburgh Department refurbished and amended many others.

The text has been updated to take account of bacterial species which have reached significance in human infection in the last few years; several colleagues have offered critical comments on the content of this volume and these are appreciated, but the errors are of my own making.

The tolerance and support of the publishing team deserve special mention; my secretary Mrs D. Burns willingly assisted in typing the manuscript but, as usual, the main burden fell upon my wife who also gave much time to proof-reading.

Belfast, 1983   R.R.G.

This volume is dedicated to the late
T. C. Dodds, FIMLT, FIIP, FRPS,
the 'father' of numerous textbooks

Contents

SECTION 1

General introduction

The four chapters in this section are aimed at silhouetting relevant elementary facts and features of the bacterial cell as a prelude to the systematic consideration of the bacterial genera associated with disease.

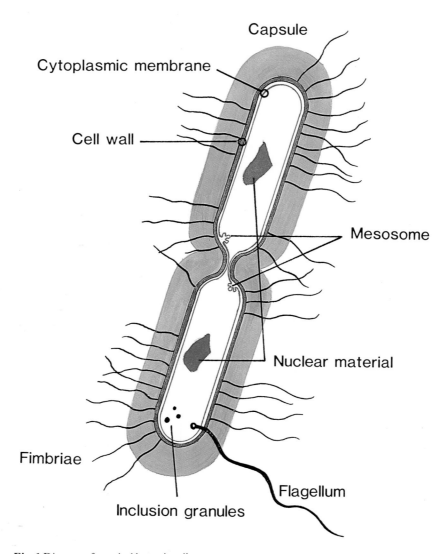

Fig. 1 Diagram of a typical bacteria cell

The bacterial cell

1

Despite their small size and their apparent simplicity when viewed in stained preparations, bacterial cells are a seething mass of biochemical activity and have been described as 'bags of enzymes'; the relevance of such physiological functions to the medical bacteriologist lies, *inter alia,* in species differentiation but the synoptic comments in this chapter are offered as a prelude to detailed classification and identification.

BACTERIAL MORPHOLOGY

The unit of measurement is the micrometre (μm), i.e. 1/1000 of a millimetre or 1/25 400 of an inch. Most cocci (spherical bacteria) are approximately 1 μm in diameter, thus 25 400 cocci would stretch across the diameter of a 2p piece!

By custom and habit textbooks continue to advise the reader of the size of various bacterial species but there is little merit in memorising such measurements and it is wiser to build up an impression of the *relative* sizes of species, e.g. the anthrax bacillus is massive (4–8 μm × 1–1.5 μm) compared with the whooping cough bacillus (1–1.8 μm × 0.3–0.5 μm).

Estimations of cell dimensions have usually been made on preparations of bacteria which have been stained after being heat killed and thus may have little relationship to the living cell. Furthermore, cell size is dependent on many environmental factors such as the age of the cell, availability of foodstuffs and the presence of antagonistic substances in the environment.

Pleomorphism

Some bacterial species show remarkable variation in size and shape and on occasion the presence of such altered morphology can lead to a tentative identification, e.g. *Haemophilus influenzae* normally presents as a small Gram-negative cocco-bacillus, but in films made from the cerebrospinal fluid in cases of *H. influenzae* meningitis the organism often appears as very long, snake-like filaments bearing terminal or lateral projecting spheres.

Pleomorphism is commonly seen in many species *in vitro* in ageing cultures or when growth takes place in the presence of antagonistic substances. It is worth noting that species which are normally Gram-positive frequently stain negatively by Gram's method when ageing or dead.

Cell wall

All bacteria are contained by a wall comprising *inter alia* a mucopeptide complex; this complex is of alternating links of two amino sugars, N-acetylglucosamine (NAGA) and N-acetylmuramic acid (NAMA) and the chains are cross-linked by pentapeptide bridges. NAMA does not occur in mammalian cells; thus the penicillin group of antibiotics can selectively damage bacterial cell walls without injury to host cells; the strength of bacterial cell walls is enormous and is reflected in the fact that they can withstand intracellular osmotic pressures of several atmospheres.

Cytoplasmic membrane

This membrane, often called the plasma membrane, lies immediately within the cell wall; it comprises mainly lipoprotein, is extremely thin and selectively allows absorption of nutrients and excretion of waste products. Its several enzymes participate in the construction of cell wall materials and in cell respiration.

Both the cell wall and cytoplasmic membrane are involved in reproduction which is by simple binary fission; after nuclear division has occurred a transverse septum grows in from the cell wall and when this septum or cross wall is complete it may split immediately to give two independent daughter cells. Splitting may be deferred so that daughter cells adhere whilst they in turn undergo further division and ultimately cells are grouped in pairs, chains or clusters depending on the relationship of consecutive planes of cleavage as daughter cells are produced from previous generations.

Mesosomes

These arise from the cytoplasmic membrane and protrude into the cytoplasm; they are seen commonly in Gram-positive species and in some Gram-negative cells. Mesosomes are visualised by electron microscopy of thin sections of bacteria and are thought to be involved in cross wall formation and also in the formation of endospores.

Nuclear apparatus

Unlike eukaryotic cells the 'nucleus' of bacteria does not possess a limiting membrane or nucleoli. It consists of irregular elongated condensations of DNA which divide by simple fission and in cells cultivated in optimal conditions *in vitro*, two, four, or even more multiples of nuclear material can be seen within a single cell as a preliminary to cross wall formation and the creation of daughter cells.

Ribosomes

Each bacterial cell contains tens of thousands of ribosomes which are minute (10–20 nm). They can be seen only with the electron microscope and contain protein and RNA; although similar to human tissue ribosomes there are distinctive differences and we can capitalise on these by interfering with bacterial metabolism at the ribosomal level without upsetting the host's ribosomes.

Intracellular granules

Various aggregations of materials can be seen in bacterial cells when suitably prepared films are viewed by the compound light microscope and their composition can often be judged by their affinity for certain dyes, e.g. lipid granules by their affinity for fat soluble dyes. A commonly occurring granule is the volutin granule, and the presence of such granules assists in the differentiation of members of the genus *Corynebacterium;* volutin granules disappear from bacteria when they are grown on mildly antagonistic media, and conversely reappear when cells are subcultured on enriched media. They are considered to be reserve sources of energy and food. Sulphur granules and granules of a starch-like polysaccharide are also found in some species but, unlike the volutin granule, these do not at present have any relevance to species differentiation.

Capsules and extracellular slime

Bacteria which form a thick gelatinous circumscribed layer outside the cell wall are described as capsulate; many saprophytic species can form capsules so that contrary to popular belief capsulation is not necessarily equated with virulence. However in species which parasitise the human host the possession of a capsule may reduce the prospect of phagocytosis; even when capsulate species are engulfed by phagocytes they may withstand digestion and multiply intracellularly with ultimate destruction of the phagocyte.

In such instances phagocytosis can be regarded as a two-edged sword!

Capsules develop rapidly when potentially capsulate species are growing in host tissues but are maintained less readily or not at all on *in vitro* cultivation. Capsular substances are usually highly specific, chemically and immunologically, and allow type differentiation of otherwise identical species, e.g. pneumococci, *H. influenzae* and *Klebsiella*.

Capsules usually are NOT revealed by ordinary staining methods and those of bacteria in cultures are most satisfactorily demonstrated with India ink. A loopful of a fluid culture or a fleck of a surface colony is emulsified in a loopful of India ink, then a coverslip is placed over the emulsion and pressed firmly down with a sheet of blotting paper. In contrast to the clearly delineated capsules, many capsulate and some non-capsulate species produce *extracellular* or loose slime; this colloidal material gives surface colonies a mucoid or sticky appearance and consistency. In India ink films, loose slime appears as irregular masses extending from the producing bacteria; in medical bacteriological practice extracellular slime is most commonly associated with *Klebsiella* species.

Flagella

With the exception of spirochaetes, motile strains of bacteria possess one or more filamentous appendages known as flagella; a flagellum is a thin filament twisted spirally, usually $0.2 \mu m$ thick and longer than the bacterial cell from which it extrudes.

The distribution of flagella (Fig. 2) is constant in any one species and peritrichous and monotrichous arrangements are the most frequent in pathogenic species; flagella cannot be observed by ordinary microscopic techniques unless they are artificially thickened by the deposition of stains (usually silver stains) on their surface. They are readily visualised by electron microscopy.

For diagnostic purposes the presence of flagella is inferred by observing motility, either in wet film preparations viewed microscopically, or by noting the spreading growth which occurs when the organism is inoculated into a tube of semi-solid agar (Fig. 3). This latter macroscopic method is preferable, since Brownian movement may be mistaken for motility in wet-film preparations, and also, the intermittency of motility implies that without repeated and frequent observations a culture may wrongly be regarded as non-motile.

Furthermore, the preparation of satisfactory wet films is time-consuming and potentially dangerous to the technician and the alternative macroscopic method should thus encourage regular examination for motility without which the identification of many species, e.g. in the enterobacteria, will be delayed.

The value of motility to the bacterial cell is not fully known but it is possible that motile species gain an advantage by spreading more rapidly through body fluids and tissues; motility may, by continuously changing the fluids in contact with the cell, assist the latter in the uptake of fresh foodstuffs and also allow the cell to move away from antagonistic environments.

Flagella have great practical importance to the bacteriologist since specific antisera can be

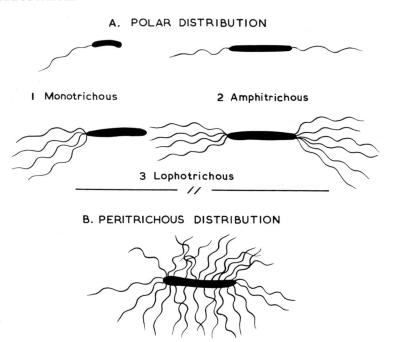

A. POLAR DISTRIBUTION

1 Monotrichous   2 Amphitrichous

3 Lophotrichous

——— // ———

B. PERITRICHOUS DISTRIBUTION

Fig. 2 Diagram of various distributions of flagella

Fig. 3 Macroscopic motility test. A pure culture is inoculated by means of a sterile straight wire into a tube of semi-solid agar to a depth of 12–25 mm. The tube on the left is uninoculated; the right-hand tube has been inoculated with a non-motile organism and after incubation growth is restricted to the inoculum track. The organisms inoculated into the middle tube were motile and they have spread throughout the medium; the inoculum track is not visible

prepared in animals and used for serological differentiation of various flagellar types; this is the basis of type-identification of the many members of the genus *Salmonella*.

Fimbriae

These are usually one half the width of flagella and are unrelated to motility; they are most commonly encountered in Gram-negative species and can be seen only by electron microscopy, but their presence can be implied by the haemagglutinating activity of most fimbriate species. Most potentially fimbriate bacilli display phase-variation, i.e. fimbriation is encouraged by serial sub-cultivation in fluid media every 28–48 h and the non-fimbriate phase appears when serial subcultures are made on solid media.

Little is known of their function but since fimbriae occur in saprophytic and commensal bacteria as well as in pathogenic species, they are probably unrelated to disease production. Haemagglutination by fimbriate species is only one example of their adhesiveness and one can readily appreciate that fimbriate strains of bowel bacteria have an advantage over non-fimbriate species since the former can adhere to intestinal epithelial cells and are thus less easily excreted.

At present the role of fimbriae in diagnostic laboratories is one of potential confusion in certain serological tests, e.g. antibodies to fimbriae are present in the sera of healthy individuals and such sera will react *non-specifically* in tests with diagnostic (stock) suspensions of salmonellae if the latter are fimbriate. Thus such tests may be falsely reported as positive.

It should be noted that several types of fimbriae can be identified using criteria such as the species of red blood cells which are aggultinated, whether or not such haemagglutination can be prevented (or reversed) by the addition of a small amount (e.g. 0.5%) D-mannose to the r.b.c. suspension and also the diameter of the fimbriae. Type 1 fimbriae are the most common and are associated with *Escherichia coli.* salmonellae and shigellae; their haemagglutinating activity is mannose sensitive whereas type 4 fimbriae haemagglutinate susceptible species in the presence of D-mannose. At present the differentiation of the various types of fimbriae has no application in routine bacteriological work.

Sex fimbriae (pili)

Long fimbriae, usually only one per bacterial cell, attach male (donor) cells to female (recipient) cells and during such conjugation DNA is transferred from donor to recipient cell.

Sex fimbriae are hollow and it is probable that the DNA passes through this core from one cell to the other.

Spores

Only two genera of medical importance, i.e. *Bacillus* and *Clostridium,* have the ability to form spores, which are analogous to air-raid shelters. Spores are highly resistant to adverse environmental conditions and as an example

of such resistance it should be noted that although all medically-important vegetative bacteria are easily killed by exposure to 70°C for 10 min, spores will happily survive such exposure for many hours and are only destroyed by autoclaving at 100°–120°C or *higher* for 10 or more minutes.

Similarly, spores are much more resistant than the parent vegetative cell to disinfectants, desiccation and sunlight, and the epidemiological significance of spores in infections caused by members of the genus *Clostridium,* e.g. tetanus and gas gangrene, and of *Bacillus* species such as *B. anthracis* and *B. cereus* is dealt with in detail later.

The enhanced resistance of spores is due to several factors including the hard spore case, their very low metabolic activity, their low content of unbound water and their high content of calcium and dipicolinic acid.

From the diagnostic viewpoint spores remain clear and unstained in filmed preparations treated by Gram's staining method, but can be demonstrated by a modification of the Ziehl-Neelsen technique although their acid-fastness is less than that of tubercle bacilli; it must also be noted that the size, shape and position of the spore relative to the 'parent' vegetative cell is constant for any one particular sporing species (Fig. 4). Thus by noting the particular combination of these three features the species can be tentatively identified, e.g. the spores of tetanus bacilli are projecting, spherical and terminal, 'the drumstick bacillus'.

A further practical use of spores lies in the checking of the efficiency of autoclaves and/or the operator; by including sealed, standardised spore populations in envelopes in trial packs within the autoclave and, after autoclaving, attempting to recover viable cells from these spore packs we can determine whether sterilisation has been effected. Strains commonly used for this purpose are *Bacillus subtilis* var. *globigii* (spores destroyed at 105°C/15 min), and *Bacillus stearothermophilus* whose spores require exposure to 121°C/20 min for destruction.

7

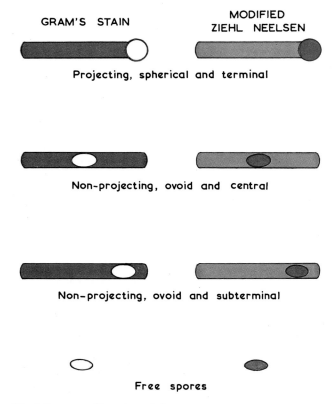

Fig. 4 Spores: position, size and shape

STAINING TECHNIQUES

Staining is of primary importance for the recognition of bacteria since their clear protoplasm is so feebly refractile that it is difficult to see them in the unstained condition unless darkground illumination is employed or, alternatively, use is made of phase-contrast techniques. It is a prerequisite that the suspension to be stained is firmly fixed to a glass slide before the various dye solutions are applied.

Making and fixing of a film

The bacterial loop is sterilised by heating to redness whilst held vertically in the flame of a Bunsen burner and then allowed to cool; the material to be filmed, e.g. a broth culture or a specimen such as sputum or urine, is then picked up by the sterile loop and spread in the centre of a cleaned microscope slide. It is important to ensure that the material being spread is kept clear of the edges of the slide to prevent manual contamination in further manipulations.

Resterilise the loop and allow the film to dry in the air; the drying process can be expedited by holding the slide *high* over the Bunsen flame.

When the film is dry the material is fixed on to the microscope slide by slowly passing the slide three times through the Bunsen flame with the slide inverted so that the coated side of the slide is in the flame (Fig. 5).

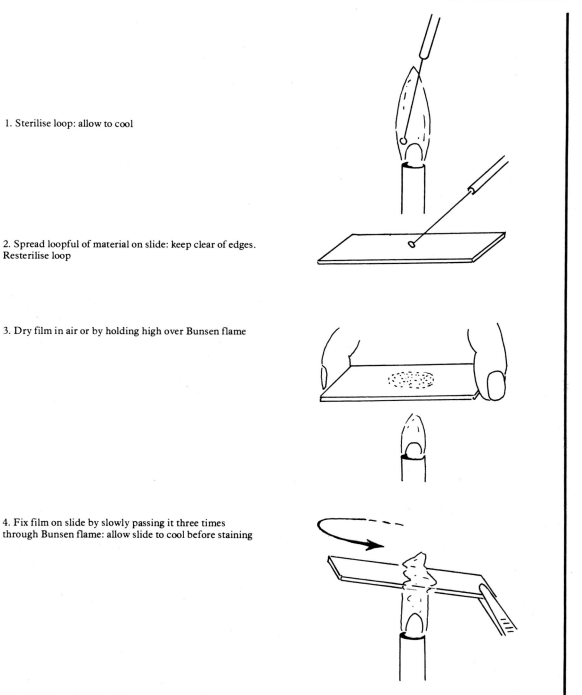

1. Sterilise loop: allow to cool

2. Spread loopful of material on slide: keep clear of edges.
Resterilise loop

3. Dry film in air or by holding high over Bunsen flame

4. Fix film on slide by slowly passing it three times
through Bunsen flame: allow slide to cool before staining

Fig. 5 Making and fixing film

Gram's staining method (Fig. 6)

This is the most commonly employed and important of all diagnostic staining procedures. After staining, bacteria are recognised as Gram-positive (purple) if they retain the primary dye complex of methyl violet and iodine in the face of attempted decolorisation *or* as Gram-negative (red) if decolorisation occurs and the bacterial cells accept the counterstain.

There are several modifications of Gram's original staining method and one such is shown in Figure 6.

1. Flood slide with methyl-violet solution. Allow to act for 5 min.

Methyl-violet stain
1% aqueous solution of methyl-violet, 6B 30 parts
5% solution of sodium bicarbonate 8 parts

2. Wash off stain with iodine solution

Iodine solution
Iodine   2 g
Normal solution of sodium hydroxide   10 ml
Distilled water   90 ml

3. Allow iodine to act for 2 min

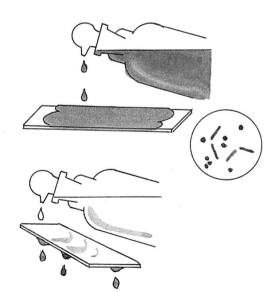

4. Drain off excess iodine. Decolorise with acetone for not more than 5 s

Fig. 6 *(see opposite)*

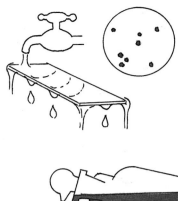

5. Wash slide immediately in water

6. Apply basic fuchsin counterstain for 30 s

Basic fuchsin stain is a 0.05% aqueous solution of basic fuchsin

7. Wash in water, blot and dry in air

The inserts indicate the appearance of a mixed Gram +ve and Gram −ve film at different stages during staining

A. Before acetone decolorisation all organisms appear Gram +ve
B. After acetone decolorisation those organisms which are Gram −ve are no longer visible
C. These Gram −ve organisms are visualised after the application of the counterstain

Fig. 6 Gram's staining method

Ziehl-Neelsen's staining method (Fig. 7)

This technique was introduced by Ehrlich to demonstrate an unusual feature of the genus *Mycobacterium,* namely acid-fastness; bacteria of this genus, as well as bacterial spores and the 'clubs' of actinomyces are relatively resistant to ordinary dyes but when they have been stained by the Ziehl-Neelsen method they resist decolorisation by strong mineral acid, 20% H_2SO_4.

Subsequent to attempted decolorisation by such an acid, and also by alcohol, the counterstain (methylene blue) is applied to give tinctorial differentiation of non-acid-alcohol-fast material.

The degree of acid-fastness varies; tubercle bacilli will retain the original red carbol-fuchsin stain when challenged with 20% H_2SO_4, leprosy bacilli only withstand decolorisation by 5% H_2SO_4, bacterial spores tolerate only 0.5% H_2SO_4 and in the attempted decolorisation of actinomyces 'clubs' the strength of acid used is 1%.

This staining method is shown pictorially in Figure 7.

1. Flood slide with carbol fuchsin. Allow to act for 5 min. HEAT intermittently without boiling the stain.

Ziehl-Neelsen carbol fuchsin
Basic fuchsin   1 g
Absolute alcohol   10 ml
Phenol solution (5% in water)   100 ml
The dye is dissolved in the alcohol and added to the phenol solution

2. Wash with water

3. Flood slide with 20% H_2SO_4. After 1 min wash in water and apply fresh acid. Repeat process several times.

4. Wash thoroughly with water

Fig. 7 *(see opposite)*

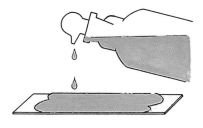

5. Apply 95% alcohol for 2 min

6. Wash with water

7. Apply methylene blue counterstain for 15 s

Loeffler's methylene blue
Saturated solution of methylene blue in alcohol   30 ml
KOH (0·01% in water)   100 ml

8. Wash in water, blot and dry in air

Since tap water may contain saprophytic acid-fast mycobacteria it should not be used to make up staining reagents for use in the Ziehl-Neelsen method; similarly washing of the preparation at all stages during staining should be with water known to be free from such saprophytic species

Fig. 7 Ziehl-Neelsen's staining method

Albert's staining method (Fig. 8)

This is a popular method of demonstrating the presence of volutin granules in various bacteria; such granules are classically associated with diphtheria bacilli and other members of the genus *Corynebacterium*. The end result of this simple staining technique is that the volutin granules stain black within a green cytoplasm. It is essential to remember that many Gram-negative bacilli possess volutin granules so that an Albert-stained film must always be evaluated alongside of Gram-stained preparation of the same bacterial population. Albert's staining method is depicted in Figure 8.

1. Apply solution 1; allow to act for 3–5 min

Solution 1
Toluidine blue   0·15 g
Malachite green   0·2 g
These are dissolved in 2 ml of 95% alcohol and added to 100 ml of distilled water containing 1 ml of glacial acetic acid. Ready for use after standing for 24 h and being filtered.

2. Wash in water; BLOT DRY

3. Apply solution 2; allow to act for 1 min

Solution 2
Iodine   2 g
Potassium iodide   3 g
Distilled water   300 ml

4. Wash and blot dry

Fig. 8 Albert's staining method

Bacterial physiology

2

Nutrition

Bacteria, as livings cells, are subject to a wide variety of environmental influences in regard to their growth and survival; the basic metabolism of bacteria is very similar to that of mammalian cells in that they require an energy source, a carbon source and a nitrogen source for the synthesis of proteins and nucleic acids and also a supply of many inorganic salts, vitamins and other accessory growth factors.

This similarity between bacterial and mammalian cell metabolism is not surprising since chemical analysis of bacterial cells has shown that their composition is very similar to that of higher animals and that the amino acids, nucleotides and fatty acids occurring in bacteria are often identical to those in higher organisms. However, subtle differences in cell wall structure, referred to in the previous chapter, do allow us to attack bacteria with antimicrobial substances without injuring the host cell.

Bacteria vary in their ability to produce complex organic substances from simple materials and many non-parasitic bacteria can utilise CO_2 as a sole source of carbon and obtain energy for synthesis from sunlight (photo-autotrophs) or by oxidation of inorganic material (chemo-autotrophs); the simplicity of such metabolic activity led one bacteriologist to describe one essentially saprophytic species as being 'the nearest thing to spontaneous generation known to man'.

Parasitic species however, cannot use such simple sources of carbon or energy and must be provided with organic nutrients such as carbohydrates, amino acids, etc (heterotrophic bacteria); in general one can state that the more parasitic an organism has become, the more exacting it is in regard to its diet and as the extreme to autotrophic bacteria, viruses are unable to synthesis their requirements and are dependent on the host cell for pre-formed foodstuffs — hence their strictly intracellular existence. Similar parallels are seen in temperature requirements and viability outside the host cell since some parasitic species, e.g. staphylococci, can not only survive but multiply outside the human host, whereas others, e.g. gonococci, cannot flourish *in vitro* for very long unless pampered in the laboratory.

Respiration

Most popular parasitic bacteria have a wide tolerance to the presence or absence of free atmospheric oxygen in their environment and grow under a wide range of oxygen tensions — the *facultative anaerobes*. Other species, e.g. tubercle bacilli, are *obligate aerobes* and demand that oxygen is freely available; at the other extreme *obligate anaerobes,* e.g. tetanus bacilli, cannot grow unless all traces of oxygen are removed from their environment and indeed the most obligate of anaerobes not only refuse to grow in the presence of oxygen, but are killed by such exposure since they then produce peroxides which are rapidly lethal in the absence of the enzyme catalase which obligate anaerobes cannot produce. This forms the basis of hyperbaric oxygen therapy for some anaerobic infections.

Certain species, e.g. campylobacters, are described as being *microaerophilic* since they grow more rapidly and luxuriantly in the presence of *traces* of free oxygen; we are increasingly aware also that many species have their growth stimulated *in vitro* when the CO_2 content of their environment is increased to 5—10% or even higher.

Temperature
Each species of bacterium has a temperature range within which growth will take place and somewhere within this 'maximum–minimum' range lies the 'optimum' at which temperature a particular species grows best (provided that all other factors are simultaneously available); the optimum temperature is usually that of the natural habitat of the bacteria, e.g. *circa* 37°C for species parasitic on man and other mammals.

Mesophilic bacteria are those that grow within the range of 25–40°C and include all species parasitic on man and certain saprophytes living freely in soil and water.

Psychrophilic species grow best below 20°C and will grow, albeit slowly, at 4°C so that although they are essentially non-pathogenic to man they have an obvious potential to affect the human race indirectly by food spoilage; this is one of the potential penalties of human progress when we rely increasingly on bulk refrigerated food storage.

The *thermophilic* groups of bacteria flourish at temperatures of 55–80°C and are non-pathogenic to man.

It must be appreciated that mesophilic species, whether or not they are capable of producing disease in mankind, are *not destroyed* by being held at refrigeration temperatures but conversely exposure of such species to temperatures significantly higher (60°C or more) than the maximum of their growth range is lethal to any mesophilic species. Thus destruction of bacteria is efficiently carried out by heating methods; under moist conditions, death is due to coagulative denaturation of the bacterial proteins whilst exposure to dry heat causes death by oxidation and charring. In general, the higher the temperature employed the shorter is the time required for sterilisation but other factors are involved in the amount of heat required to destroy bacteria; they are more readily killed in the presence of acids or alkalis and less readily killed if non-bacterial proteinaceous or other organic materials are present. Thus it is essential that equipment to be sterilised should firstly be thoroughly cleaned.

Thermal death point. This (TDP) is defined as the lowest temperature above the maximum at which growth occurs at which a given species is killed in a specified period of time (usually 10 min); for most non-sporing mesophilic species the TDP is 60°C on exposure to moist heat and between 100–120°C for most sporing species. The saprophytic species *Bacillus stearothermophilus* requires moist heat at 121°C for 10–30 min to ensure destruction of its spores and is used as one biological indicator that sterility has been obtained in autoclaving of surgical dressings, etc.

D value. The D value or *decimal reduction time* is a *practical* index of the heat sensitivity or heat resistance of bacterial species; the D value is the time of exposure (in minutes) at a given temperature that results in a ten-fold reduction in numbers of a test bacterial suspension under standard conditions and appears to be more reliable than the TDP.

Hydrogen-ion concentration

There is a wide variation in the tolerance range of pathogenic bacteria, e.g. pH 4.0–9.0, although bacteria that parasitise man prefer a restricted pH range of 7.2–7.6. However, cholera vibrios grow best at pH 8 and are intolerant of acid environments; on the other hand some organisms are *acidophilic*, e.g. lactobacilli, and flourish in a pH of less than 4.

Obviously we have to cater for such species when attempting *in vitro* isolation by providing artificial media suitably adjusted to their requirements, and similarly environmental alterations of pH may help to destroy species in nature.

Influence of moisture

More than 80% by weight of the bacterial cell consists of water and as in higher organisms, moisture is essential not only for growth but for the survival of vegetative cells. Drying under natural conditions is tolerated differently by various species, and tubercle bacilli may survive for weeks or months after leaving the human host whereas gonococci survive only briefly in nature.

Bacterial spores are very resistant to desiccation as indicated by the survival of anthrax spores for more than 80 years when dried on to linen threads.

Desiccation *per se* need not necessarily be lethal even to the most delicate species and this is exemplified by our use of lyophilisation methods, where by combining *rapid* freezing and drying techniques, such freeze-dried pure preparations of bacteria can be preserved for years without subculture.

Light and other radiations

Virtually all parasitic bacteria grow and survive best in darkness and ultra-violet rays are rapidly lethal whether derived from an artificial source or naturally, either as direct sunlight or as 'sky-shine', i.e. diffuse sunlight.

Radiations are used to sterilise certain modern medical materials and radiation from a Cobalt-60 source is routinely used to ensure that disposable plastic syringes and Petri dishes are sterile before use in the wards or the laboratory.

CULTIVATION OF BACTERIA

There are only a few instances when one can categorically identify a bacterium solely by the examination of microscopic specimens so that routinely we attempt to harvest the causal organism of a disease on artificial culture media. Such nutrient media are dispensed as fluid in test tubes or screw-capped bottles or in Petri dishes if solid; solidification of fluid media is easily effected by adding a small amount of agar, an inert polysaccharide derived from seaweed.

Nutrient broth is the basis of many media used in diagnostic laboratories and is a watery solution of peptone and meat extracts; peptone is a mixture of polypeptides and amino acids serving as sources of nitrogen, carbon and energy and is obtained by digesting meat with trypsin or other proteolytic enzymes. Meat extract, comprising the water-soluble components of meat, supplements peptone by providing a variety of mineral salts and growth factors. Nutrient agar is the solid equivalent of the broth to which has been added 1–2% agar.

Enriched media have added materials, e.g. 5% blood to give a blood agar medium to enhance growth of more exacting organisms.

Selective media can be prepared by adding a substance or substances which depress the growth of some species without interfering with others, e.g. the addition of a minute amount of crystal violet to blood agar allows the growth of streptococci whilst inhibiting the growth of staphylococci.

Indicator media have evolved to allow macroscopic recognition of chemical reactions; one of the earliest of these was MacConkey's medium containing, *inter alia,* a small amount of lactose and an indicator of pH change so that species capable of growing on the MacConkey agar and also utilising lactose (with the production of acid) produce red-coloured colonies whereas lactose non-fermenting species growing on the same medium give rise to colourless colonies.

Examples of these and other media will be given as each bacterial species is dealt with in the section on Systematic Bacteriology.

Plating out

Isolation of pure cultures is essential for the identification of bacteria and is normally achieved by plating out the specimen directly on to the surface of a solid medium in a Petri dish; the inoculation on to such a solid medium may also be derived from fluid media in which the specimen has been initially incubated.

When plating out, a well inoculum area is spread homogeneously over a quarter of the surface of the solid medium using a wire loop loaded with pus or fluid culture or using a swab of material inoculated from the infected lesion.

The wire loop is then sterilised in a Bunsen flame and, after cooling, is recharged from the well inoculum and drawn in two or three parallel lines across a fresh part of the medium, this procedure is repeated using as inoculum the most distal part of the immediately preceding strokes (Fig. 9A). Thus the inoculum density is successively reduced so that ultimately, isolated organisms are deposited and on incubation these organisms (colony-forming units) give rise to individual, separated colonies.

Each colony is a *pure culture* of a single kind

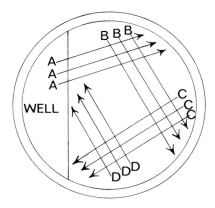

A

B

Fig. 9 A. Normal method of plating out. A reducing inoculum from the well area through the series of strokes is ensured by sterilising the inoculating loop at each stage. **B.** Plating out on selective media. Such media can be more heavily inoculated and there is no need to sterilise the inoculating loop between each series of strokes

of bacterium since it consists exclusively of the progeny of a single bacterial cell.

In circumstances where it is known that mixtures of bacterial species will be present in a specimen, e.g. faeces, it is essential to use selective media to inhibit the growth of commensal organisms thus allowing any pathogenic bacteria (which may be scanty) to obtain more nutrient; in the case of faecal specimens this is effected by using a deoxycholate citrate agar (DCA) plate; such selective media can be more heavily inoculated and there is no need to flame the inoculating loop between successive strokes (Fig. 9B).

Also in such specimens, enrichment media can be used in parallel with the primary selective medium; fluid enrichment media, e.g. selenite broth and tetrathionate broth improve the isolation rate of some popular intestinal pathogens such as salmonellae and shigellae and these are harvested from the enrichment broth by subinoculation to fresh plates of DCA.

Having provided adequate nutrients, most bacteria will grow in the natural atmosphere but special apparatus is employed to grow obligate anaerobes.

Incubation

Inoculated media are maintained at the optimum temperature (usually 37°C) by use of a thermostatically controlled incubator or warm room which is dark and thus excludes harmful ultra violet radiations; if incubation is prolonged then the mouths of test tubes containing culture medium must be sealed with tight fitting rubber caps or alternatively the medium can be dispensed in screw-capped bottles.

Fig. 10 Anaerobic jar with lid *in situ*

ANAEROBIC CULTIVATION

The realisation that some bacteria would grow only in the absence of atmospheric oxygen led to various methods of obtaining anaerobic conditions which culminated in the production of an anaerobic jar by McIntosh and Fildes in 1916. The original jar and method of use have been modified significantly in recent years, and provided that the seal between the jar and its lid are airtight and the jar is properly used, even the most exacting anaerobic species can be isolated (Fig. 10).

It is important that vented Petri dishes should be used to allow free access of the anaerobic atmosphere and that the culture medium used should be freshly prepared; the inoculated plates are placed in the jar in an inverted position (the culture containing portion uppermost) and the lid secured on to the jar by the screw clamp.

The jar is then evacuated to 630–660 mmHg below atmospheric pressure and then a mixture of 90% hydrogen with 10% CO_2 is admitted to equilibrate.

After 10 min a check is made to ensure that a secondary vacuum has developed from the cold catalyst which is suspended from the interior of the jar lid.

Finally the jar contents are equilibrated with the hydrogen/CO_2 gas mixture before incubation; the cold catalyst consists of alumina pellets coated with palladium and are contained in a fine gauze sachet. It is important that the catalyst is reactivated frequently by drying in a hot air oven.

Hydrogen sulphide is produced during cultivation of most anaerobic species and this is the main agent which poisons the palladium

...yst; a recently introduced commerical ...roduct, Anotox, can adsorb H_2S as well as the volatile fatty acids produced by growth of anaerobic organisms. Thus the inclusion of a packet of Anotox in the anaerobic jar prolongs the life of the catalyst.

'Gas Pak' method

The introduction of cold catalyst eliminated the need for 'sparking' with electricity as in the original jar and more recently commercial packs have become available for the generation of hydrogen and CO_2 so that bulky cylinders of the gas mixture are no longer necessary. The commercial pack is made active simply by adding 10 ml of water and then placing the opened pack upright in the anaerobic jar which is immediately sealed in the usual manner (Fig. 11).

Fig. 11 The simplicity of the 'GasPak' system of generating hydrogen and carbon dioxide is shown here; the foil envelope is opened by peeling back the corner to a printed line and 10 ml of water is injected. The envelope is immediately placed upright in the anaerobic jar, the lid replaced and screwed down and the jar is then ready for incubation

Indicators of anaerobiosis

Earlier chemical methods of demonstrating the absence of oxygen in the incubating anaerobic jar have been superseded by placing a culture plate seeded with the obligate aerobe *Pseudomonas pyocyanea* in the jar; no growth will result on this test plate if anaerobic conditions are acquired and maintained.

Of course the bacteriologist frequently requires to culture many specimens in which anaerobic bacteria might be present alone or in company of other organisms and for this purpose a cooked meat broth can be inoculated and incubated without recourse to anaerobic jars. Such a medium, preferably freshly made in the laboratory, consists of cooked, minced, sterile ox heart muscle, and this contains reducing sustances which maintain anaerobic conditions in the depths of the medium and will also support the growth of most aerobes in the nutrient supernatant layer. Ultimate isolation of organisms from inoculated cooked meat broth requires subculture to solid media incubated anaerobically and aerobically.

Bacterial growth curves (Fig. 12)

The laboratory study of bacterial growth is one of increasing interest, particularly if continuous culture systems are employed which ensure a steady maintenance not only of fresh suitable nutrient but of all other physico-chemical conditions permitting optimum growth of the particular bacterial species; similarly the concurrent growth of two bacterial species in such continuous culture systems is pursued in the hope of elucidating symbiotic and antibiotic phenomena.

At the more mundane, but equally important, level the clinician requires a knowledge of factors influencing bacterial growth (*and* survival) when for example he is required to submit specimens of urine for examination within strict time limits after the specimen has been collected; alternatively he must be made aware of the feeble viability of many species, particularly the anaerobes, and specimens which might contain such species

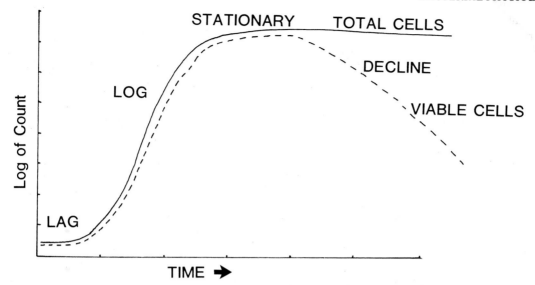

Fig. 12 Bacterial growth curve

require speedy transmission to the laboratory, or protection from atmospheric oxygen, otherwise false negative culture results will be reported.

When bacteria are inoculated into a *fresh* fluid medium there is a *lag phase* of some duration before active multiplication can be detected; the duration of this lag phase varies with different bacterial species and also on the nature of the fresh medium. If the latter is identical with the medium from which the inoculum was derived the lag phase is short since there is no need for the bacteria to adapt its enzymatic activities to different nutrients; although there is little or no *multiplication* of cells during the lag phase (indeed some cells may die) one can demonstrate increases in metabolic activity and cell size.

Towards the end of the lag phase there is a rapid increase in the rate of cell division and the population then enters the *logarithmic phase* of growth when, under optimal conditions, most ordinary species, such as the pyogenic cocci and members of the enterobacteria, show an exponential increase in numbers and there is a straight line relationship between the log of the number of cells plotted against time. This logarithmic

phase ends in a closed culture system when nutrients become exhausted and simultaneously the accumulation of toxic by-products of growth begin to promote autolysis.

During the *stationary phase* the number of cells remains constant and there is no further growth; this phase is variable in duration and eventually the *decline phase* is reached and here the predominant feature of the population is one of increasingly rapid death and/or autolysis.

The differences between *total* bacterial counts (which includes dead cells) and the *viable* count are most obvious in the decline phase but if autolysis is a dominant feature the *total* count will not greatly exceed the viable number of cells; as mentioned earlier, dead cells which were originally Gram-positive often stain Gram-negatively and pleomorphism is also a feature of cells in the decline phase.

It should be obvious that these various phases also occur in human tissues and fluids so that specimens of urine must be delivered promptly to the laboratory (ideally within one hour of collection) and if the period is likely to exceed 2–3 h then falsely high bacterial counts will be found. Methods of combating such false positive findings will be dealt with in a later section.

Antigens and antibodies

3

Physically detectable reactions between bacteria and the serum from patients recovering from infection were noted by the founders of present-day bacteriology and this field of interaction between antigens and antibodies has exploded in a most spectacular fashion in the last few decades; one of the compelling reasons for this dramatic expansion of immunological techniques is the need for tissue-typing of recipients and donors of transplanted organs, but the more mundane uses of immunological reactions continue. In summary, it is a daily requirement to give a detailed identification of many species of bacteria which are identical in their morphology, cultural characteristics and biochemical reactions and such detailed indentification is frequently undertaken by identifying highly specific antigens in the isolated bacteria using laboratory derived specific antisera.

Likewise there are many occasions when it is not possible to isolate bacteria from an infected patient, perhaps because he is being investigated in a late stage of infection, but *indirect* evidence of infection can be obtained by detecting in a sample of his blood serum antibodies specific to a laboratory-maintained stock culture of a particular microorganism.

Antigens

An antigen is *any* substance which stimulates the host's immune system to form antibody and when mixed with that antibody the antigen reacts *specifically* with it in some observable way.

Bacteria and their products are antigenic and in common with other antigens they are usually proteins although non-protein substances, such as the polysaccharides comprising pneumococcal capsular material, are also efficient antigens.

Antibodies

An antibody is a globulin protein that appears in the blood serum and tissue fluids of a host in response to the introduction of an antigen, and which when mixed with that antigen reacts specifically with it in some observable way.

The different serological reactions that can be demonstrated with an antiserum are largely determined by the physical state of the antigen employed in the test system; thus we speak of:

Precipitation. When the bacterial antigen is present in colloidal solution and is layered carefully over its specific antiserum, precipitation occurs at the interface; the most common use for such tube precipitation tests is in determining the group-specificity of β-haemolytic streptococci (Fig. 13).

Alternatively, by incorporating the

Fig. 13 Precipitation test. In diagnostic laboratories, precipitation tests are most frequently employed in determining the group-specificity of β-haemolytic streptococci; the method of extracting the group-specific carbohydrate antigen is shown in Figure 34. In this example the left-hand tube contained group A antiserum and that on the right, group C antiserum; equal volumes of antigenic extract were then carefully superimposed on each antiserum. Within 5 min the tube on the left showed heavy precipitation at the interface whilst that on the right remained unaltered. Therefore, the strain belonged to group A, i.e., *Strept. pyogenes;* another method of rapidly identifying these strains of β-haemolytic streptococci which are pathogenic to man is shown in Figure 104A

antiserum in an agar plate of medium and then introducing the soluble antigen, various bands of precipitate may be noted since different antigens diffuse at different rates and one can thus determine the number of components in a mixture of antigens. This gel-diffusion method of demonstrating precipitation is primarily associated with the Elek or Ouchterlony method of demonstrating the presence of diphtheria exotoxin diffusing from a streak culture of diphtheria bacilli interacting with specific antitoxin on an impregnated sheet of blotting paper placed at right angles to the bacterial growth; lines of precipitate can be noted somewhere in the angle between the line of bacterial growth and the antitoxin impregnated paper where the toxin and its antitoxin meet in suitable proportions (see Fig. 49 on p.   )

More sophisticated and rapid techniques

based on *precipitation* reactions are now in use, e.g. immunoelectrophoresis, where the antiserum is placed in a small well cut in the surface of agar borne on a glass slide; the application of a direct electric current allows differential migration of the serum antigen components to occur from the serum along the agar gel; these are then demonstrated by placing, in a thin trough cut longitudinally in the agar gel, antiserum against the electrophoresed antigen and the two components diffuse towards each other with the formation of various bands of precipitate which can be clearly visualised by applying a protein stain.

Agglutination. Here the antigen consists of a suspension of intact bacteria which clump together in the presence of specific antiserum and these clumps aggregate and settle as a deposit with clearing of the supernatant. Such reactions can be made quantitative by employing serial (doubling) dilutions of the antiserum in a set of tubes to each of which is added a constant aliquot of antigen; thus the agglutinating titre of a serum is stated as the reciprocal of the highest serum dilution with clearly visible agglutination (Fig. 14). This is the basis of the Widal test to detect salmonella antibodies.

Agglutination tests are frequently carried out on a microscope slide as a preliminary step in the serological identification of many organisms, e.g. shigellae; here a loopful of a saline suspension of the 'unknown' bacterial culture is mixed with a loopful of stock antiserum and where there is specificity between bacterial antigens and the stock serum aggregation occurs. In non-specific mixtures the bacterial suspension remains homogeneous (Fig. 15).

Opsonisation. By estimating the number of bacteria ingested by phagocytes in the presence of a patient's serum in comparison with the number phagocytosed in a non-immune serum an index of opsonic activity can be calculated, e.g. if on counting the bacteria within 50 phagocytes in the presence of a patient's serum the average per phagocyte

23

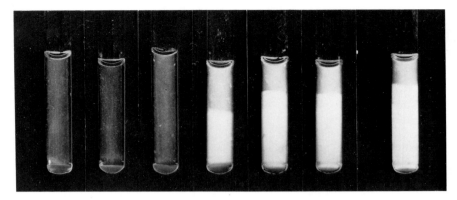

Fig. 14 Tube agglutination test. Tubes 1–6 contain doubling dilutions of a patient's serum (30–960) and tube 7 is a control tube without serum which shows that the bacterial suspension is not autoagglutinable. All seven tubes had an equal volume of bacterial suspension *(S. paratyphi B)* added and were then incubated in a water-bath at 37°C for 4 h. The tubes were then removed from the bath and left at room temperature for 2 h.

Deposition of the agglutinated bacteria, with complete clearing of the supernatant, has occurred in tubes 1–3; the agglutinating titre of the serum was therefore 120.

In tubes 4–6, and in the control tube, agglutination has not occurred (see caption to Figure 102 for details of the technique)

Fig. 15 Slide agglutination tests. These are employed as a preliminary step in the identification of many organisms. In this instance loopfuls of a saline suspension of a bacterial culture (? salmonella) were placed separately on the slide; a loopful of group A *Salmonella* antiserum was added on the left and on the right, a loopful of group B *Salmonella* antiserum was similarly mixed with the bacterial suspension. The slide was rocked gently and within 30 s agglutination was obvious in the mixture on the right; on the left the bacterial suspension was unaffected and remained homogeneous. Subsequent detailed examination revealed that the organism was *S. typhimurium*. The results of slide agglutination tests must always be confirmed by tube agglutination tests

is 10 whereas in non-immune serum only 2 bacteria per phagocyte are noted, the opsonic index of the patient's serum equals 5. Such tests are technically difficult to standardise and statistically unreliable and are not now in routine use.

Neutralisation. If a bacterium or its toxin is lethal for a laboratory animal then the co-existence of its antibody or antitoxin should be protective to the animal.

In determining whether diphtheria bacilli isolated from a patient are toxin-producing (i.e. virulent), two biologically equivalent guinea-pigs are injected with aliquots of a culture of the bacilli *but* one of the animals will previously have been protected with diphtheria antitoxin. If the strain of diphtheria bacillus is virulent then the protected animal will remain healthy and the unprotected guinea-pig will show characteristic pathological changes; the absence of pathological changes in the latter animal indicates that the organism is avirulent (or that it is not a diphtheria bacillus).

Obviously, and as in all other antigen-antibody reactions, neutralisation tests can also be used to detect antibody if one uses a known antigen; similarly this and other serological reactions can be rendered quantitative by using a series of animals or test-tubes and maintaining a constant concentration of one reagent whilst varying the concentration of the other.

Bacteriolysis. Intact bacterial cells are disrupted in the presence of specific antibody and serum complement.

By themselves, neither antibody nor complement can effect dissolution of bacteria.

Complement is a group of proteins present in the *normal* (non-immune) serum of many animals which deteriorates rapidly *in vitro* unless kept in the frozen state and can be inactivated by heating serum to 55°C for 30 min, which treatment does *not* affect the antibody content.

Haemolysis of red blood cells (r.b.c.) by a haemoyltic antiserum is analogous to bacteriolysis, thus:

RBC + haemoyltic serum + complement ⟶ haemolysis

RBC + haemolytic serum (heated to 55°C/ 30 min) ⟶ no haemolysis

RBC + complement ⟶ no haemolysis

Thus we have, by suspending red cells in *heated* haemolytic serum, a 'sensitised' or indicator system which, when added to another mixture, will either remain unaltered if complement is absent from that mixture or will show a greater or lesser degree of haemolysis if complement is present.

Complement fixation. As in all other antigen-antibody reactions complement-fixation (Fig. 16) can be used to detect antibodies by using a stock antigen or conversely to detect antigens by employing stock antiserum. Although complement-fixation was originally developed by Wassermann for the detection of syphilitic antibodies, the technique is now more widely applied in other, particularly virological, areas.

Fig. 16 The result of a *quantitative* complement-fixation test performed with a positive serum, i.e., one which contained antibody specific for the test antigen.

Equal and constant volumes of fresh complement and stock antigen are in each tube with doubling dilutions of patient's serum which had been heated at 55°C for 30 min to destroy its natural complement; the dilution in the left tube is 1 in 5, in the next tube 1 in 10 and in the tube on the right the serum dilution is 1 in 20.

After primary incubation for 1 h at 37°C one volume of sensitised r.b.c. was added to each tube, the contents thoroughly mixed and the tubes replaced in the water bath at 37°C for 30 min.

In the first tube no haemolysis has occurred since there was sufficient antibody to bind all of the complement and none was available to complete the indicator system comprising the r.b.c. with its heated specific haemolytic antiserum.

In the middle tube, although most of the complement was fixed by the primary antigen-antibody reaction the residue allowed partial haemolysis of the sensitised r.b.c.

In the third tube, although antigen and antibody united, there was insufficient antibody to fix most of the complement and enough of the latter was left unbound to effect complete lysis of the sensitised r.b.c.

If this test had been performed with a patient's serum which was negative, i.e., did not possess antibody specific for the test antigen, the result in all three tubes would have been identical and similar to that of the third tube

The technique depends on two separation reactions. Firstly, antigen and antibody (one of which is unknown) are mixed with a predetermined amount of fresh complement — the antibody-containing serum having been heated to destroy its natural complement. If antigen and antibody are specific for each other then their union will use up or 'fix' the fresh complement; evidence of fixation can only be determined by adding a second 'sensitised' system of r.b.c. suspended in their heated specific haemolytic antiserum. The absence of free complement in the primary reactions allows the red cells to remain intact; conversely if the primary antigen and antibody are unrelated, they will not unite and the added complement will remain free and available to complete the sensitised system, when it is added, with resultant lysis of the red cells.

In complement fixation tests, therefore, lysis of the red cells in the sensitised system indicates that the patient's serum in the primary reaction did *not* contain antibody for the particular antigen employed; the absence of lysis shows that the patient's serum contained specific antibody for the antigen, i.e. the patient's serum was 'positive'.

When the bacteriologist is asked for a decision on the significance of the level of antibody present in a *single* specimen of patient's serum he requires full information regarding the particular patient, e.g. duration of illness, if and when a particular vaccine has been given, and even then it is unusual to state with confidence that the determination of the presence of specific antibody at any given titre can be equated with infection caused by the organism against which antibody has been detected.

Interpretation of the significance of antibody titres is greatly enhanced when two separate serum specimens are submitted, one early in the illness and the second specimen some 7–10 days later; the demonstration of a *significant rise in titre* between the first (acute phase) serum and the second (convalescent phase) serum can almost be taken as evidence of infection. A significant rise in titre is accepted when there is a four-fold rise in the level of antibody from the acute to the convalescent phase serum.

All too often however the patient's infection has progressed before a serum sample is submitted for evaluation and the titre to a particular antigen will then probably already be at a high level; if the causal organism cannot be detected then a retrospective diagnosis may be obtained if one can demonstrate a *fall* in antibody titre in a serum sample taken some *weeks* after the first specimen.

Classification and identification

4

Lower bacteria

These are unicellular, never form a mycelium and each cell in the clone is biologically independent; lower bacteria are much more numerous than higher bacteria and have much greater significance as human pathogens.

Higher bacteria

Filament formation is the rule in these which often show true branching with the formation of a mycelium; interdependence within the clone is a regular feature, e.g. some cells being specialised for reproduction. The filaments of higher bacteria are often sheathed.

A broad classification of the *lower bacteria* is based simply on cell shape (Fig. 17), and we speak of

Cocci. The cells are spherical or occasionally very slightly elongated.

Bacilli. Relatively straight, rod-shaped and non-flexuous cells.

Vibrios. These rigid cells are definitely curved in appearance.

Spirilla. Cells resemble a corkscrew, spiralled but non-flexuous.

Spirochaetes. Cells are highly flexuous, spiralled filaments; the majority of

spirochaetes have a very fine cross diameter and cannot be seen microscopically by ordinary techniques.

Each of these morphologically distinguishable groups of lower bacteria is further subdivided.

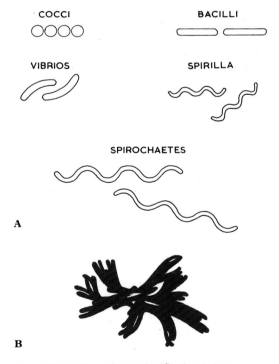

A

B

Fig. 17 Bacteria — primary classification **A.** Lower (eubacteria) **B.** Higher (actinomycetales)

Cocci

These are readily differentiated on the basis of their reaction to Gram's staining method *and* the relationship of consecutive planes of cleavage as daughter cells are produced (Fig. 18); of the medically significant cocci all stain Gram-positively except members of the Neisseriae and these Gram-negative cells adhere mainly in pairs; the cells are slightly elongate at right angles to the axis of each pair (line-abreast formation) and the opposed surfaces are flat or sometimes concave.

Staphylococci. Gram-positive cells which adhere in grape-like clusters, since consecutive planes of division are haphazard.

Streptococci. These adhere in chains, since successive cell divisions occur in the same plane as earlier cleavages; Gram-positive.

Diplococci. These Gram-positive cells adhere in pairs or in very short chains; in fresh specimens and young cultures the cells in each pair are slightly elongate in the axis of the pair (line-ahead formation).

Gaffkyae. Gram-positive cells which appear as flat plates of four cells, since consecutive planes of division alternate at right angles.

Sarcinae. These are seen as cubes or packets of eight cells due to division occurring successively in *three* planes at right angles; Gram-positive.

Bacilli

Unfortunately these cannot be subdivided in as straightforward a manner as the cocci and various characteristics have to be used even for elementary differentiation (Fig. 19); certain bacilli possess the feature of *acid-fastness* when stained by Ziehl-Neelsen's method.

Non acid-fast bacilli can be divided into those which stain positively by Gram's method and Gram-negative genera.

A simple subdivision of Gram-positive bacilli is made on the basis of spore formation, and spore-forming Gram-positive bacilli can be further subdivided into two large groups depending on whether the vegetative cells are aerobic in their atmospheric requirements, i.e. the genus *Bacillus,* or anaerobic, i.e. the genus *Clostridium.*

Non-sporing Gram-positive bacilli comprise many genera which can be recognised and differentiated only by detailed study of their biochemical activities and serological make up.

The third and final group of bacilli are those which stain Gram-negatively, and again a broad subdivision of Gram-negative bacilli is made on their atmospheric requirements for growth. The anaerobic Gram-negative group is characterised by *Bacteroides* species and the aerobic Gram-negative group by *Pseudomonas* species.

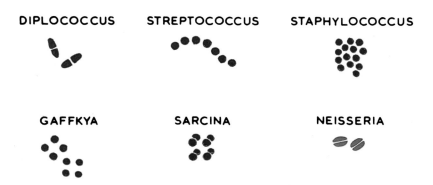

Fig. 18 Classification of cocci

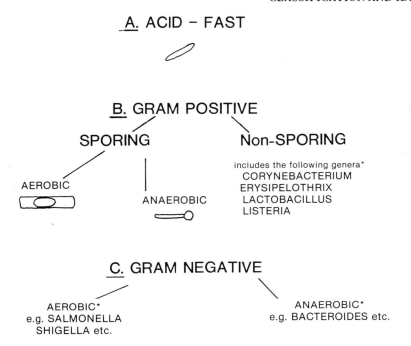

A. ACID – FAST

B. GRAM POSITIVE

SPORING   Non-SPORING

AEROBIC   ANAEROBIC   includes the following genera*
CORYNEBACTERIUM
ERYSIPELOTHRIX
LACTOBACILLUS
LISTERIA

C. GRAM NEGATIVE

AEROBIC*   ANAEROBIC*
e.g. SALMONELLA   e.g. BACTEROIDES etc.
SHIGELLA etc.

*can be recognised and differentiated only by further study

Fig. 19 Classification of bacilli

Within each of these groups of Gram-negative bacilli, recognition of genera requires further investigation of cultural, biochemical and other characteristics.

Vibrios and spirilla

All members are Gram-negative and mostly motile, having polar flagella; differentiation of species is by cultural, biochemical and serological methods.

Pathogenic spirochaetes

Three genera embrace all spirochaetes associated with infection in man. *Borrelia* are Gram-negative, larger than other pathogenic species and can be stained and viewed microscopically by ordinary methods. *Treponema* are much finer and possess more coils than *Borrelia;* they are only

BORRELIA

TREPONEMA

LEPTOSPIRA

Fig. 20 Classification of spirochaetes

demonstrable by dark-ground microscopy or if stained by a silver impregnation method to increase their size artificially. *Leptospira* are even finer in structure than treponemes and possess more numerous coils which are so close to each other as to be barely discernible; one or both ends are recurved on the body of the organism (Fig. 20).

Motility is endowed not by flagella but by a series of axial filaments wound round the protoplast and fixed at both poles of the spirochaete; these filaments impart motility by rapid alternating contraction and expansion.

DIAGNOSTIC IDENTIFICATION OF BACTERIA

There are many occasions when detailed identification of a species is sought, either for epidemiological purposes or because it is rare and bacteriologists are, or should be, curious. However the clinician is primarily interested in knowing whether an isolate is causing infection, so that the bacteriologist has to temper his laboratory curiosity with the need for speedy and accurate identification; the following outline of laboratory procedures can often be short-circuited because of unique features of certain bacterial species.

Microscopy

Using simple staining methods applied directly to pathological material or to pure cultures isolated from specimens we can place bacteria in their appropriate group depending on their morphology and staining reactions. In some instances this otherwise preliminary step in identification is sufficient in itself, e.g. the presence of acid-and-alcohol-fast bacilli in sputum (tubercle bacilli) and the presence of *intracellular* Gram-negative diplococci in cerebrospinal fluid (meningococci).

In a few instances, e.g. leprosy and Vincent's infection, microscopy offers the only means of identification, since the causal organisms cannot at present be grown.

Microscopic characteristics which should be noted routinely are the size, shape and arrangement of cells to each other, the response to Gram's staining method and the presence or absence of motility, capsules and spores; likewise any special staining features should be noted, e.g. the presence of volutin granules.

Cultural requirements and appearances

These are a prerequisite in differentiating many species, e.g. within the enterobacteria, which are microscopically identical. Colonial characteristics which should be noted include the size of the colony, its shape in plan view and elevation (Fig. 21), opacity or transluscence, pigmentation, either natural, as in staphylococci, or resulting from indicator

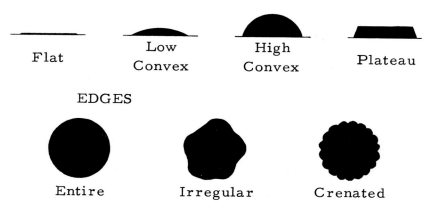

ELEVATIONS

Flat   Low Convex   High Convex   Plateau

EDGES

Entire   Irregular   Crenated

Fig. 21 Elevations and edges of common types of colony

changes in differential media, e.g. *Escherichia coli* giving pink colonies on MacConkey's medium due to acid production from the fermentation of lactose.

The consistency of the colony and any changes in the colour or consistency of the underlying medium should be noted, e.g. haemolysis on blood agar; details of such characteristics are given in following chapters.

Biochemical reactions

In many bacterial genera, commensal and pathogenic species are not only microscopically identical but yield, in pure culture, colonies which are not clearly distinguishable. The biochemical activities of such pure cultures frequently allow species differentiation and fermentation reactions are popularly employed for this purpose (Fig. 22).

Fermentation reactions. Fermentation of a series of carbohydrates, or other substrates, is tested by growing the organism in a fluid medium containing a small quantity of the test sugar (usually 1%) and an indicator which reacts tinctorially if acid is produced; an inverted miniature test tube (Durham tube) is immersed in the medium and gas formation resulting from substrate utilisation is revealed by the collection of bubbles at the

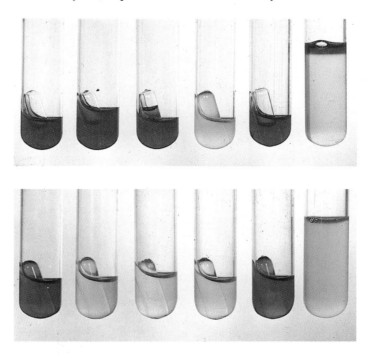

Fig. 22 Biochemical differentiation of microscopically similar organisms. 1% solutions of glucose, lactose, dulcitol, sucrose and mannitol in peptone water are contained respectively in tubes 1–5; the sixth tube contains plain peptone water.

The upper row of tubes was inoculated with an organism (*Shigella sonnei*) which ferments only glucose and mannitol without producing gas. Indole is not produced in the peptone water culture.

The lower row of tubes was inoculated with an organism (*Escherichia coli*) which, with gas production, utilised all substrates except sucrose and also produced indole. The production of indole from peptone is readily tested by adding 1 ml of ether to the peptone water culture which is then shaken vigorously, and, after allowing the tube to stand for 2 min, 0·5 ml of Ehrlich's rosindole reagent is added

31

apex of the Durham tube.

Thus by noting the varying patterns of substrate utilisation and whether or not gas is produced one can recognise various groupings, e.g. in the enterobacteria, salmonellae and shigellae are readily differentiated since the former almost always produce gas (and are motile) whereas the latter rarely produce gas and no *Shigella* species is motile.

Other enzymatic activities. Various tests will be mentioned in the section dealing with different bacterial genera and whilst some, e.g. the oxidase test, have only an indentificatory value, others, e.g. the coagulase test, allow the recognition of pathogenic species.

Serological reactions

These have already been referred to in Chapter 3 and are often the basis of detailed identification of an organism by detecting highly specific antigens using laboratory-prepared antisera; such analysis is very gratifying to the bacteriologist and allows him to relate serotypes isolated from different patients to each other (e.g. in an outbreak of salmonella food poisoning) but such details are secondary to more basic information, namely that the isolate is 'some salmonella or other', in the management of each individual patient.

Animal inoculation

The use of experimental laboratory animals is restricted as far as possible on humanitarian grounds and also because of the expense of breeding and maintenance; however in a few instances animal inoculation has to be resorted to either for recovery of a particular pathogen, e.g. *Leptospira interrogans* serotypes or the differentiation of species, e.g. human type tubercle bacilli do not cause lesions in rabbits whereas tubercle bacilli of the bovine type cause progressive infection in rabbits.

The different epidemiology of human and bovine type tubercle bacilli in mankind requires us to know which type of bacillus is causing infection in a given case. This allows the tracing of the source of infection and its eradication as a continuing method of spreading infection in the human population.

Epidemiological markers

Brief reference has been made to some techniques, e.g. serotyping of species, which allow detailed marking or tagging to permit identity of isolates from different sources to be established.

Other methods of highly specific identification include *bacteriophage typing:* bacteriophages are viruses which are parasitic on bacteria and display highly specific relationships with their host cell, and like all viruses they multiply only within the host cell. Phages are able to lyse many of their bacterial hosts so that by implanting selected phage preparations on a culture plate seeded with a culture of the susceptible bacteria one can, after incubation, note activity of the phage preparation macroscopically by the presence of plaques or zones of lysis. Different phage activity patterns allow the recognition of phage types. The method is used to type pathogenic staphylococci, several of the more commonly encountered salmonellae, e.g. *Salmonella typhimurium,* and more recently phage typing of Group B β-haemolytic streptococci has also been established as a means of type identification.

In addition to serotyping and bacteriophage typing of various bacterial species another epidemiological marking system depends on *bacteriocine production;* bacteriocines are naturally occurring antibiotic substances whose activity is usually restricted to other species within the same genus as the bacteriocine producing strain.

Bacteriocine typing is effectively used to subdivide a species which is serologically homogeneous, e.g. *Shigella sonnei,* and the technique is simple; the strain to be typed (the producer strain) is streaked diametrically across the surface of a suitable medium which is then incubated under strict conditions regarding temperature and time. During incubation any bacteriocines produced will diffuse into the medium and their presence is ultimately detected by streaking out a stock set of indicator strains at right angles to the original growth line of the producer strain.

Reincubation then reveals which of the indicator strains is inhibited by the bacteriocines and depending on the pattern of inhibition we recognise various bacteriocine types of producer strains.

SECTION 2
Systematic bacteriology

In this section salient details are given to allow differentiation of pathogenic members of each genus from each other and from essentially commensal members of that genus.

The various tables of differential characteristics are intended to exemplify the use of biochemical and other tests in the recognition of species within a given genus; IT IS NOT THE AUTHOR'S INTENTION THAT THESE SHOULD BE MEMORISED.

The following key can be used for all these tables:

Fermentation reactions:
$\perp$ = Acid production without gas formation
+ = Acid and gas produced
– = No fermentation
( ) = Delayed reaction

Other reactions:

Gelatin liquefaction
+ = Liquefaction occurs
– = No liquefaction

Indole production
+ = Indole produced
– = No indole produced

Motility
+ = Motile
– = Non-motile

Growth, e.g. on MacConkey's medium
+ = Growth occurs
– = No growth

The abbreviation 'V' in any table indicates that strains within a group are variable, e.g. some *Shigella flexneri* strains produce indole and other strains do not.

Staphylococci

5

It is just over 100 years since Ogston discovered staphylococci, which are ubiquitous and can be isolated from innumerable living hosts including man as well as from air, dust, water and foodstuffs. Staphylococci form a dominant part of the commensal flora of our skin and mucous surfaces and at the same time are capable of causing infections varying from the trivial, localised, superficial skin lesion to severe and often lethal septicaemia.

Microscopy

Spherical, approximately 1 μm in diameter, grape-like clusters and Gram-positive; non-motile, non-sporing, capsules detectable in some freshly isolated pathogenic strains (Fig. 23).

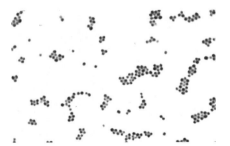

Fig. 23 Gram-stained film of pure culture of staphylococci
$\times$ 1000

Cultural appearances

Grow abundantly on all ordinary media, facultatively anaerobic; colonies have an entire edge and a convex elevation showing golden or white pigmentation and are 2–4 mm in diameter after 18–24 h incubation at 37°C (Fig. 24). Pigmentation is less marked when incubated anaerobically but pigment develops

Fig. 24 Blood agar plate inoculated from a nasal swab. After 18 h incubation at 37°C a pure and profuse growth of *Staph. pyogenes* var. *aureus* was evident. These proved to be coagulase +ve, indicating that the individual from whom the swab was taken was a carrier of pathogenic staphylococci. The size of these colonies should be contrasted with the much smaller (and non-pigmented) colonies of streptococci. (See Figs. 31 and 35)

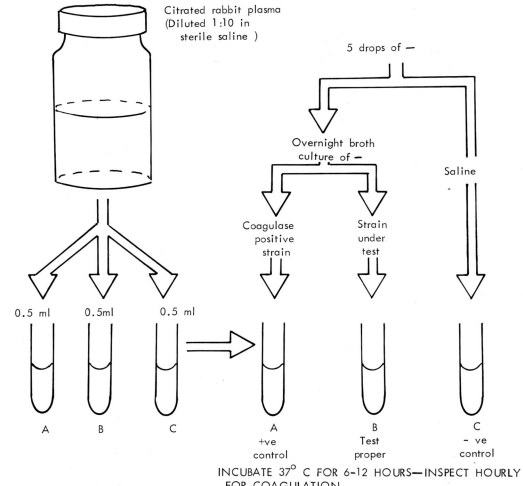

Citrated rabbit plasma
(Diluted 1:10 in
sterile saline)

5 drops of —

Overnight broth
culture of —

Saline

Coagulase
positive
strain

Strain
under
test

0.5 ml   0.5ml   0.5 ml

A   B   C   A   B   C

A   +ve   control

B   Test   proper

C   – ve   control

INCUBATE 37° C FOR 6-12 HOURS—INSPECT HOURLY
FOR COAGULATION

Fig. 25 Coagulase test

when such cultures are exposed to air; pigmentation is enhanced by cultivation on milk or cream agar but nowadays no reliance is placed on particular pigment production as a guide to pathogenicity.

Biochemical reactions

The *coagulase test* (Fig. 25) has, for years, proved the most reliable and convenient test to differentiate the popular pathogenic strains which we call *Staph. pyogenes*.

Coagulase is a prothrombin-like enzyme

produced by more than 98% of staphylococci isolated from human infections; the *tube* coagulase test detects free enzyme and the alternative method of coagulase detection, the *slide* test, relies on the presence of bound coagulase.

In the slide test a smooth suspension of the staphylococcal colony being investigated is made in a drop of normal saline; then a drop of undiluted, normal rabbit plasma is mixed with the bacterial suspension. Coagulase positive staphylococci will be clumped within 15 s as they bind together, because fibrin is

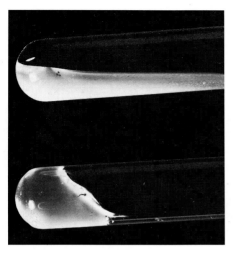

Fig. 26 Coagulase tests after incubation at 37°C for 6 h. The upper tube shows a negative test; the contents are fluid. In the lower tube the citrated plasma has been coagulated, indicating a positive result

precipitated on the cell surfaces (Fig. 26). Simultaneous control tests using known positive and negative reactors are undertaken.

The *phosphatase* test may also be used to detect pathogenic staphylococci; the strain is grown on an agar medium containing phenolphthalein diphosphate and if phosphatase is produced, free phenolphthalein is liberated in the colony and when the latter is exposed to ammonia vapour the colony turns a bright pink colour.

The phosphatase test is not as widely used as the coagulase test but is certainly useful when searching for carriers of pathogenic staphylococci since swabs, e.g. from the anterior nares, can be plated directly on to phenolphthalein diphosphate agar plates.

Staph. pyogenes colonies usually produce a golden pigment (var. *aureus*), occasionally a white pigment (var. *albus*) and more rarely the colonies show lemon pigmentation (var. *citreus*).

Until a few years ago coagulase negative strains were equated with a commensal status and grouped together under the title *Staph. epidermidis* reflecting their constant presence on skin and mucosal surfaces of healthy individuals; the realisation that certain

coagulase negative staphylococci are associated with certain infections, e.g. of cerebrospinal fluid shunts in hydrocephalics, has stimulated tremendous interest in this hitherto ignored group of staphylococci.

At present, nine species of coagulase negative staphylococci are recognised and two of these, *Staph. epidermidis* and *Staph. saprophyticus,* predominate as potential pathogens; the former produces phosphatase and is sensitive to the antibiotic novobiocin whereas *Staph. saprophyticus* strains are phosphatase negative and resistant to novobiocin.

Serological characters

Although agglutination and precipitation tests can be used to serotype *Staph. pyogenes,* the information thus obtained in epidemiological studies of staphylococcal infections has been less satisfactory than that obtained by phage typing.

Phage typing (Fig. 27) The method is shown diagramatically in Figure 28 and is

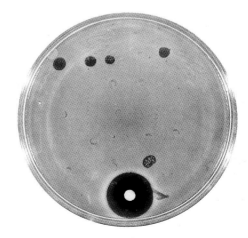

Fig. 27 Phage typing of staphylococci. A plate of digest agar has been marked with a grid of 22 squares, each corresponding to the site of a particular phage preparation. The surface of the plate was flooded with a broth culture of a staphylococcus and then allowed to dry. The phage preparations were then individually applied each in 0·01 ml amounts and when these had dried the plate was incubated overnight at 37°C. On examination confluent lysis was noted in the areas of phage preparations 3B/3C/55/71

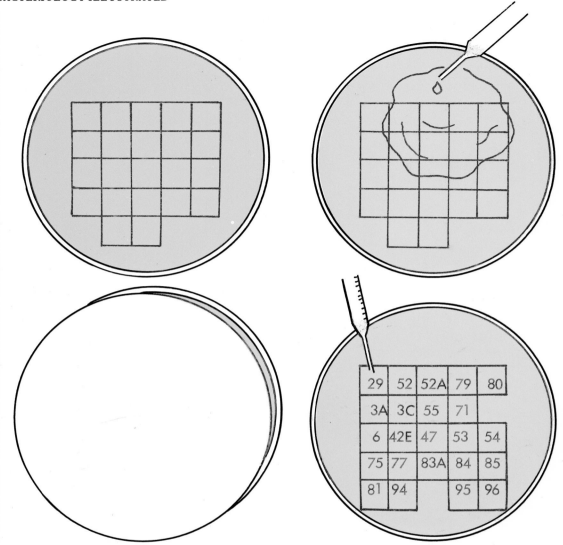

Fig. 28 Phage typing technique **1.** Grid plate of digest agar **2.** Surface flooded with overnight broth culture of strain to be typed **3.** Lid replaced incompletely. Left at room temperature for 1–3 h **4.** 0.01 ml drop of each typing phage applied. Plate allowed to dry — incubated at 37°C for 12–18 h

invaluable in unravelling epidemics of *Staph. pyogenes* infection, e.g. in post-operative wound infections. The phage typing of isolates from the wounds, from carriers, from air samples and fomites, etc will reveal the source and spread of the epidemic strain in the environment and on this evidence measures can be instituted to limit if not terminate the epidemic.

Animal inoculation

Such procedures have no role in the routine laboratory but animal experiments play an important role in the study of several toxins produced by pathogenic staphylococci.

STAPHYLOCOCCAL INFECTIONS

Clinically, staphylococcal infections are associated with pus formation and circumscribed lesions and these associated with *Staph. pyogenes* species can be classified as *superficial,* e.g. boils, carbuncles and wound infections, or *deep;* of the deep-seated staphylococcal infections acute osteomyelitis, particularly of the long bones, was once among the most dreaded of diseases.

In the pre-antibiotic era, even with speedy diagnosis, osteomyelitis frequently had a chronic course with the need for intermittent surgical removal of bone sequestra and with sinus formation. Now with early diagnosis, isolation of the causal organism and prompt antibiotic therapy based on *in vitro* antibiogram testing, few cases progress to the chronic stage which often required limb amputation to save the patient from the slow but inevitable outcome of amyloid disease.

Other deep infections include pneumonia, perirenal abscess and acute endocarditis.

Staphylococcal food poisoning

Some *Staph. pyogenes* strains produce enterotoxins when growing in contaminated food stuffs. When such contaminated food is ingested an acute food poisoning syndrome may result within a very few hours; the patients present in a varying state of collapse with giddiness and often furious vomiting and diarrhoea. The enterotoxins causing this *intoxication* are relatively heat stable and may survive cooking which will kill the staphylococci which produced the enterotoxin; six antigenically distinct staphylococcal enterotoxins have been indentified. The sources of such food poisoning types of staphylococci are human cases, e.g. the butcher with a hand injury infected with *Staph. pyogenes* and carriers of enterotoxin producing strains.

Toxic shock syndrome

In 1978 an apparently new staphylococcal disease was reported and at first was mainly associated with the use of tampons in menstruating women. The cardinal symptoms are hypotension, fever, erythematous rash (followed by desquamation) accompanied by evidence of toxicity in several body systems, e.g. diarrhoea, vomiting, myalgia, etc; the aetiology of this syndrome is unfolding at present and the occurrence of toxic shock syndrome in males and in non-menstruating females of all age-groups tends to detract from an early hypothesis that one particular type of tampon included material favouring the growth of staphylococci.

At present two pyrogenic staphylococcal exotoxins (PEA and PEB) are thought to be implicated in this syndrome.

Infections with coagulase negative staphylococci

These have been referred to briefly already and although many such infections occur in compromised patients, e.g. those with prosthetic heart valves and hip joint prostheses, *Staph. saprophyticus* is undoubtedly incriminated as an important causal agent in urinary tract infection in children and young women and its virulence in this situation is associated with adhesion of *Staph. saprophyticus* to the urogenital epithelium. Such infections have led to a revival of interest in the entire genus but we should not be surprised at the abilities of staphylococci to adapt to various situations since they showed, many years ago, an ability probably unrivalled by any other genus to outwit our antibiotic armamentaria.

The hospital staphylococcus

This term was coined for strains of *Staph.*

pyogenes which rapidly adapted to the hospital environment in being able not only to survive but to multiply in hospitals on any moist surface, e.g. soap, to develop resistance to penicillin G and many of its derivatives and to show tolerance to many disinfectants and antiseptics in daily use. Similarly they can tolerate salt concentrations which are inhibitory to most other pathogens and their adaptability is further confirmed by the fact that carrier rates for staphylococci are higher among hospital personnel than for people living in the open community; they may at present take second place to Gram-negative bacilli in the league of hospital-acquired pathogens but they remain as a constant and dangerous threat to patients of all ages who are hospitalised.

Streptococci

6

Streptococci are as ubiquitous as staphylococci and have a wide variety of hosts; some are members of the normal human flora and others cause human diseases directly attributable to infection or indirectly by sensitisation to them.

Microscopy

Regardless of the varying cultural, biochemical and serological characteristics of streptococci *members of this genus are microscopically indistinguishable.* Approximately 1 μm in diameter, spherical, arranged in chains of varying length, Gram-positive, non-motile, capsulate in certain circumstances and non-sporing (Fig. 29).

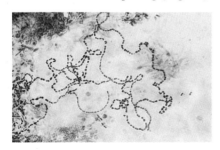

Fig. 29 Gram-stained film of *Streptococcus pyogenes* from pathological material. This should be viewed in contrast with Figure 23. Note that the chain length is no guide to the cultural type of streptococcus.
$\times$ 600

Cultural appearances

Some members of the genus are obligate anaerobes and these are dealt with at the end of this chapter. Streptococci are more fastidious than staphylococci in their cultural requirements; they grow best on blood agar which also allows a primary differentiation of the *aerobic* streptococci (Fig. 30). Three

Fig. 30 The organisms seen in the upper half of this blood agar plate show β-haemolysis and those in the lower part portray the much less distinct nature of α-haemolysis. The organisms are respectively *Strept. pyogenes* (type 4) and *Strept. viridans*

responses can be seen surrounding streptococcal colonies in blood agar BUT the colonies themselves are very similar in size and shape — they are only half the diameter (1–2 mm) of staphylococcal colonies after 18–24 h incubation at 37°C and are non-pigmented.

α-haemolytic streptococci are surrounded by a narrow halo of greenish-grey discolouration in which only a few red cells are lysed when the medium is examined microscopically.

β-haemolytic streptococci are surrounded by a very much wider zone of complete clearing of the red cells which are entirely disrupted; β-haemolysis is enhanced when culture plates are incubated anaerobically.

γ-haemolytic (non-haemolytic) streptococci produce no visible alteration in the blood agar.

These cultural types of streptococci are also clearly differentiated by their pathogenicity and communicability in the human host; α-haemolytic types (*Streptococcus viridans*) are

essentially commensal in the healthy upper respiratory tract. γ-haemolytic types (*Strept. faecalis; Enterococcus*) lead a commensal existence in the intestine of man and animals.

In certain circumstances α and γ types can assume a pathogenic role in the human host and such infections are *endogenous* in nature; by contrast, β-haemolytic streptococci account for most of the primary streptococcal infections in man and *exogenous* infection is the rule, i.e. β-haemolytic streptococci spread

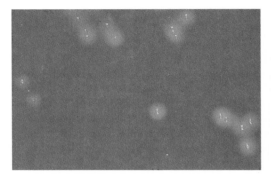

Fig. 32 An enlarged view of one area of the above plate reveals the characteristics of streptococcal colonies which, regardless of the lytic changes effected in blood agar, are similar for α, β and γ types of streptococci

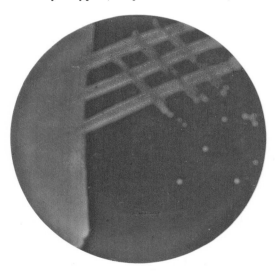

Fig. 31 Blood agar plate with a 1 in 500 000 concentration of crystal-violet incorporated. The plate was inoculated from a throat swab taken from a patient with acute tonsillitis. After 18 h incubation at 37°C a pure growth of β-haemolytic streptococci was evident; these proved to belong to Lancefield group A, type 12. The addition of crystal violet reduces greatly the growth of many other organisms, e.g., staphylococci

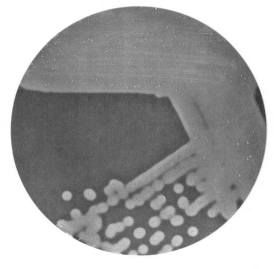

Fig. 33 This plate, inoculated from the same throat swab, was incubated anaerobically. The enhancement of β-haemolysis under these conditions is obvious

readily from one host to another in epidemic fashion.

β-haemolytic streptococci can tolerate the presence of small concentrations of crystal violet which are inhibitory to staphylococci and the incorporation of 1/500 000 parts of this dye into a blood agar plate renders the medium selective when inoculated with material, e.g. a nasal swab, containing a mixture of organisms (Figs. 31–33).

β-HAEMOLYTIC STREPTOCOCCI

Biochemical reactions

Biochemical subdivision has been attempted in an endeavour to establish epidemiologically significant types of β-haemolytic streptococci; this was not satisfactory and serological grouping and typing methods are available and reliable.

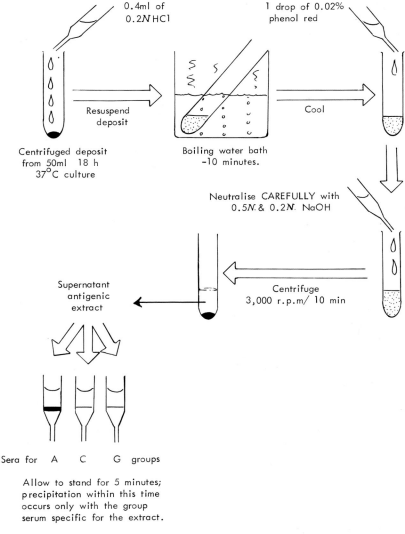

Fig. 34 Group identification of β-haemolytic streptococci: Lancefield's method

Serological characteristics

A broad grouping of β-haemolytic strains is possible by precipitation techniques in which the group-specific carbohydrate (C) antigen extracted from the organism is tested against standard group-specific antisera (Fig. 34). The groups A to S (no groups designated I or J) are in general related to different animal hosts and the majority of strains from man belong to group A.

Streptococcus pyogenes. Since group-A strains are so prevalent in human infections they have been specifically named *Streptococcus pyogenes* and have been subjected to vigorous serological analysis which now allows the recognition of many serotypes based on the content of various specific protein (M and T) antigens; strains may possess both, either or neither of these type-specific antigens. Typing of *Strept. pyogenes* is a two-step procedure — firstly, intact cells are tested by slide agglutination with antisera against various T-antigens revealing the presence of one or more T-antigens in the particular strain; the second stage uses acid-extracted M-antigen (the process is identical with that which extracts the group or C-antigen) tested for precipitation with highly type-specific M-antisera.

Strains possessing type-specific M-antigen give matt-surfaced colonies and are virulent compared with the same strain lacking M-antigen; in man, recovery from group-A streptococcal infection is associated with long-lasting immunity against the particular M-serotype of the infecting strain but no protection against the other serotypes within the group. Thus, in theory at least, one individual could suffer many separate attacks from group-A streptococci, each attack caused by a previously unencountered serotype!

Streptococcus agalactiae. This species belongs to Lancefield's group B and, unlike *Strept. pyogenes,* grows on bile-containing media. Strains vary in their haemolytic activity when grown on blood agar but, uniquely, colonies of *Strept. agalactiae* produce red or orange pigment when incubated *anaerobically* on starch-containing media.

Intense recent interest in the pathogenicity of this species has resulted in a serotyping scheme allowing the recognition of eight distinctive types and even more recently a phage typing method has evolved analagous to that used in epidemiological studies of staphylococcal infection.

α-HAEMOLYTIC STREPTOCOCCI

Such strains are usually termed *Streptococcus viridans* and have already been noted for their essentially commensal existance in the upper respiratory tract; species identification is being evolved based on a range of biochemical tests, but such taxonomic studies require further elucidation.

γ-HAEMOLYTIC STREPTOCOCCI

These strains are synonymously termed *Streptococcus faecalis* and *Enterococcus* and are commensal in the gut. Microscopically they are slightly ovoid in shape but no stress can be put on this for differentiation from other streptococci.

Biochemical reactions

Strept. faecalis can be typed biochemically and, like group B streptococci they grow in the presence of bile salts and on MacConkey's medium appear as minute (0.5–1 mm) magenta coloured colonies; they also grow in high concentrations (6.5%) of NaCl and most are tolerant of high pH environments.

Although most faecal streptococci are without effect on blood agar they belong to group D in Lancefield's classification (Table 1).

STREPTOCOCCAL INFECTIONS

Streptococcus pyogenes infections

Strept. pyogenes is readily spread from cases of infection and carriers to other susceptible

Table 1. Group D Streptococci: biochemical subdivision

Type	Sorbitol	Arabinose	Gelatin liquefaction	Growth at pH 9.6
Strept. faecalis				
var. *faecalis*	⊥	–	–	+·
var. *liquefaciens**	⊥	–	+	+
var. *zymogenes*†	⊥	–	+	+
Strept. faecium	–	⊥	–	+
Strept. durans	–	–	–	–
Strept. bovis	–	⊥	–	–

*No haemolysis on blood agar plate.
†β-haemolysis on blood agar plate.

individuals; such exogenously acquired infections are particularly common in younger people, especially school children, and include streptococcal sore throat, scarlet fever, otitis media, erysipelas, and some cases of impetigo. *Strept. pyogenes* also infects wounds and burns; many cases of puerperal sepsis were formerly caused by such strains but they are less commonly incriminated nowadays.

Infection with any of the specific serotypes of *Strept. pyogenes* may be followed by acute rheumatic sequelae and this causal relationship has been established not only on clinical grounds but also on the basis of bacteriological and serological studies. The most convincing proof of this relationship, however, was the demonstration of the remarkable prophylactic benefits accruing from long-term administration of suitable antimicrobial agents to known rheumatic subjects and similarly the fact that rheumatic sequelae do not occur if the primary streptococcal infection is adequately treated, i.e. the prompt administration of penicillin in proper dosage for a sufficient period, e.g. one mega unit of depot penicillin given by intramuscular injection will maintain adequate blood-levels for at least ten days.

Certain serotypes of *Strept. pyogenes*, particularly type 12 strains, may be nephrotoxic; thus acute glomerulo-nephritis may result from infection. It appears that even with rapid and adequate treatment of the primary streptococcal lesion the prevention of glomerulo-nephritis is not guaranteed.

Streptococcus agalactiae infections

Infection with this species has been well recognised as mastitis in cattle and the reduction in milk yield resulted in the specific epithet for the species; bovine mastitis caused by this and other organisms can have an obvious effect on man's economy.

In recent years these group B β-haemolytic strains have become directly important to mankind as a cause of neonatal and puerperal infection; *Strept. agalactiae* appear to be commensal in the large intestine and thus colonise the vulva and vagina; neonates may be colonised during birth but those delivered by caesarian section can also become colonised by manual transmission from mothers and attendant staff. Neonatal infection with this species is more commonly reported from the USA than from Britain but in both countries infection is either of the 'early-onset' septicaemic type occuring within 24 h of birth or 'delayed-onset' meningitis usually occurring in the first few weeks after birth. Although the causal organism is sensitive to the penicillins the high mortality rate is accentuated in premature babies and where complications have arisen at childbirth.

Streptococcus viridans infections

Such species have already been noted for their essentially commensal existence in the upper respiratory tract; however, in an individual with a history of rheumatic carditis (usually mitral stenosis) or a congenitally abnormal heart valve (usually a bicuspid aortic valve), *Strept. viridans* is the commonest cause of subacute bacterial endocarditis. This condition was invariably fatal before the advent of antibiotics and even yet carries a 20–35% fatality rate; infection is *endogenous,* the source of the organisms being the mouth in a state of poor dental hygiene. Other organisms are involved in about 5% of cases of subacute bacterial endocarditis and it is essential that blood culture be performed, and repeatedly, so that the causal organism can be isolated and its sensitivity to antibiotics determined as a guide to rational therapy (Fig. 35).

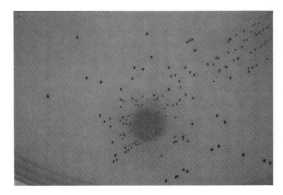

Fig. 36 MacConkey's plate showing *Strept. faecalis* colonies as minute and magenta coloured. Their size can be contrasted with those of the larger colonies of *Escherichia coli.*
× 3

Strept. faecalis strains are however incriminated in a proportion of endogenously acquired cases of urinary tract infections and often in association with *Escherichia coli* strains (Fig. 36).

Likewise they can participate as a cause of peritonitis following abdominal tragedies, e.g. a ruptured appendix; very occasionally they are incriminated causally in subacute bacterial endocarditis.

ANAEROBIC MICROCOCCI

Despite years of study there is still no consensus of opinion on the role of this group of organisms in human infections; some investigators regard anaerobic micrococci as commensals. They are also found as saprophytes and can be isolated from various sites in healthy individuals, e.g. the mouth, genito-urinary tract and skin surface. Equally anaerobic cocci are, on occasion, the only bacteria isolated from infected sites, e.g. brain abscesses.

More frequently anaerobic cocci are found in association with other species particularly those of the genus *Bacteroides.*

Equally the classification of anaerobic cocci is an area of conflict nationally and internationally so that only guide lines are

Fig. 35 Blood agar plate inoculated from a swab taken from the skin. After overnight incubation at 37°C it can be noted that two species were present: the white pigmented colonies are those of staphylococci which were coagulase –ve, and α-haemolytic colonies which were shown to be those of *Strept. viridans.* The much smaller size of the streptococcal colonies can be noted since, even including the surrounding lytic zone, the area occupied by the colonies of *Strept. viridans* just equals the size of the staphylococcal colonies

Streptococcus faecalis infections

Provided that these remain in their normal commensal habitat in the healthy large intestine they, and their host, remain content.

offered; anaerobic micrococci are significantly smaller than their aerobic counterparts and show variation in their reaction to Gram's staining method although they are usually demonstrable as Gram-positive and occur in chains of varying length (Fig. 37). Thus, at present, they are referred to as *Peptostreptococci* and/or *Peptococci*.

Cultivation requires strict anaerobic conditions on blood agar with 0.01% sodium oleate incorporated in the medium; even after 48 h incubation at 37°C colonies are small (1–2 mm) and have no specific characters allowing ready identification.

One can hope that this presently nondescript group of bacteria will receive more attention in the near future so that we can unravel their classification and, equally important, their epidemiological significance.

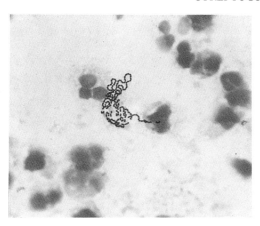

Fig. 37 Gram-stained film showing *Pepto. putridus* and polymorphonuclear leucocytes; the film was made from a high vaginal swab from a patient with puerperal sepsis × 750

Pneumococci

7

The pneumococcus *(Streptococcus pneumoniae)* is a normal inhabitant of the human upper respiratory tract. It can cause pneumonia (usually of the lobar type), paranasal sinusitis, otitis and meningitis which is usually secondary to some other pneumococcal infection.

Microscopy
1 μm in their long axis, slightly elongated and arranged in pairs when a lanceolate shape may be seen or in very short chains in the line of the long axes, Gram-positive, non-motile, capsulate and non-sporing (Fig. 38).

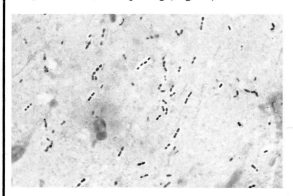

Fig. 38 Gram-stained film of sputum from a case of lobar pneumonia; pneumococci appear mainly in pairs. *N.B.* The capsules are not demonstrated by Gram's staining method.
$\times 800$

The above description holds for films made from pathological material or in freshly isolated cultures from such material; on continued laboratory cultivation pneumococci are microscopically indistinguishable from streptococci, particularly in the disappearance of their capsules.

Cultural appearances

Facultatively anaerobic but growth is enhanced by increasing CO_2 concentration to 15%; as fastidious as streptococci and produce a zone of α-haemolysis on blood agar, hence differentiation from commensal *Strept. viridans* is essential. With both organisms α-haemolysis is accentuated when grown on chocolate blood agar (Fig. 39). Colonies are plateau in elevation after 18–24 h incubation at 37°C and on continued incubation autolysis of the centre of the colony is noted – the *draughtsman colony.*

Biochemical reactions

These are of interest only in so far as they allow differentiation from *Strept. viridans;* the majority of strains ferment inulin as compared with these streptococci but more specific tests are available.

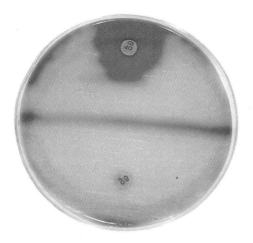

Fig. 39 A chocolate blood agar plate was sown on its upper half with a pneumococcal culture and on the lower half with *Strept. viridans.* Filter paper discs, impregnated with a 1 in 4000 solution of Optochin, were placed in each area; after 18 h incubation at 37°C inhibition of the pneumococci contrasts with the unhibited growth of *Strept. viridans.* The frequent co-existence of potentially pathogenic pneumococci and essentially commensal *Strept. viridans* in specimens from the respiratory tract demands an easy and reliable means of differentiating them. Optochin sensitivity testing is much more reliable than inulin fermentation and more readily performed than bile solubility tests.

The accentuation of α-haemolysis when such species are grown on such a medium is obvious

pneumococci are extremely sensitive and viridans streptococci are resistant; the test is performed by placing a filter paper disc, impregnated with a 1 in 4000 aqueous solution of Optochin, on the surface of a medium which has been sown with the organism.

Serological characters

At least 83 specific serotypes of pneumococci can be identified; specificity is endowed by the specific capsular polysaccharides. Typing is performed on a slide by mixing a loopful of broth culture or saline suspension from a blood agar plate culture with a loopful of diagnostic antiserum; a cover slip is applied over the mixture and the preparation viewed with an oil-immersion objective under reduced illumination.

In the presence of its specific antiserum the capsule is sharply outlined and appears swollen in comparison with parallel preparations containing heterologous type sera. In the absence of specific antibody the capsule and its margin are quite invisible (Fig. 40). In practice, strains are first tested in nine pooled sera and then in the type-specific sera

Bile solubility test. Pneumococci are soluble in bile whereas *Strept. viridans* remain intact. This feature can be tested by adding one part of a sterilised 10% solution of sodium taurocholate in normal saline to 10 parts of a broth culture. Incubation of the mixture for 15 min at 37°C will reveal whether lysis occurs; it is essential that the broth culture should be adjusted so that its pH is between 7 and 7.5 otherwise chemical precipitation of the bile salt may occur with resultant turbidity.

Optochin sensitivity. Optochin is an alkaloid, allied to quinine, to which

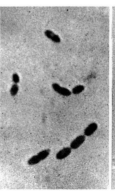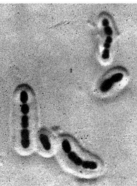

Fig. 40 The organism in both preparations was *Pneumococcus* type 19; in the left-hand plate the organism was in the presence of a *heterologous* antiserum and methylene blue in a wet film. No capsule 'swelling'. In the right-hand plate *homologous* (type 19) antiserum was present and a capsule 'swelling' reaction is obvious
× 2475

comprising the pooled serum in which a reaction was obtained; such typing was an essential preliminary in the pre-sulphonamide era when serum therapy was employed in the treatment of pneumococcal lobar pneumonia; nowadays serotyping of pneumococci is only of epidemiological interest.

PNEUMOCOCCAL INFECTIONS

Pneumococci are associated quite indelibly, and rightly so, with **lobar pneumonia** in which they are the most common species encountered; however, pneumococci are also commonly involved as secondary pathogens in cases of *bronchopneumonia* where the primary disease is viral in origin, e.g. measles and influenza.

They are also frequently encountered alone or in association with other pyogenic organisms in cases of paranasal sinusitis and acute otitis media; in such cases infection may have been acquired exogenously or alternatively the pneumococci have extended from the naso-pharynx where they have been existing as commensals. Pneumococci frequently spread from the initial focus of infection, e.g. pneumococcal meningitis occurs as a complication of otitis media and lobar pneumonia, but some cases of pneumococcal meningitis present as a primary infection. When such patients suffer recurrent attacks diligent investigation frequently reveals that there is a skull defect; such defects may be congenital, e.g. absence of a cribriform plate, or post-traumatic. Closure of the defect with a fascial graft gives a high degree of protection against recurrences.

Pneumococcal meningitis carries an unusually high mortality rate which increases with the age of the patient, and it is difficult to account for this difference *vis à vis* meningococcal meningitis since both organisms are (at present) equally sensitive to penicillin, which is the drug of choice.

Pneumococci have taken on a new significance in being a very common cause of infections in patients lacking a spleen, either from surgical removal or as part of the result of total body irradiation for therapeutic purposes; thus a polyvalent pneumococcal vaccine has been introduced to immunise such patients actively against infection with the 14 most commonly encountered serotypes; the efficacy of such immunisation awaits proof and splenectomised patients should probably be put on a long-term penicillin prophylactic regime, e.g. 1.5 mega/units of benzathine penicillin G given i.m. monthly. Controlled trials will ultimately give the answer to the efficacy of such prophylactic procedures.

Until 1976 it had been assumed that pneumococci were eminently sensitive to penicillin but in that year reports from two separate centres in South Africa revealed that penicillin resistance had emerged and the isolation of such strains has since been reported from other countries.

Neisseriae

8

Most members of this genus are commensal in the upper respiratory tract and occur extra-cellularly; the two pathogenic species (*Neisseria meningitidis,* the meningococcus, and *N. gonorrhoeae,* the gonococcus) are typically *intracellular* (Fig. 41).

Microscopy

Approximately 1 μm in diameter, slightly elongate and arranged in pairs with the long

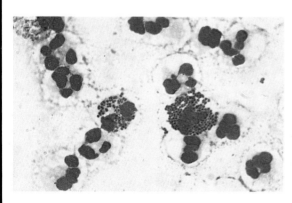

Fig. 41 A Gram-stained film of the urethral discharge from an acute case of gonorrhoea in a male. Two polymorphonuclear leucocytes are crammed with Gram-negative diplococci; this intracellular appearance is typical of the pathogenic neisseriae. This preparation could equally well be of cerbrospinal fluid from a case of meningococcal meningitis
$\times 800$

axes parallel and flattening or concavity of their opposed surfaces. Gram-negative, non-motile, non-capsulate and non-sporing; however, when strains of the pathogenic species are freshly isolated, capsules may be seen, particularly if specific antiserum preparations are used.

The shape and arrangement of the pathogenic species are usually changed on subcultivation in the laboratory; they become more spherical and tend to appear in clusters. Some strains of *N. gonorrhoeae* possess fine fimbriae.

Cultural appearances

The pathogenic species are most fastidious and growth is enhanced by incubating in an atmosphere of 10% CO_2; blood agar or heated blood agar *must* be used for primary isolation and incubation at 37°C must be continued for 48 h before one rejects a culture as negative. Selective media have been available for some time and at present one such, the modified New York City medium (MNYC) gives earlier and more luxuriant growth of *N. gonorrhoeae* than other media; MNYC is not only a rich medium but incorporates antibiotics which render it selective by suppressing the growth of contaminant organisms.

Colonies are convex, 1–2 mm in diameter, transparent, non-pigmented and non-haemolytic; considerable variation occurs and prolonged cultivation gives a crenated colony, often with opacity of the central part of the colony and radial striations.

Biochemical activities

ALL neisseriae give a positive *oxidase reaction;* this is tested by applying a freshly prepared 1% solution of the oxidase reagent (tetramethyl-*p*-phenylenediamine

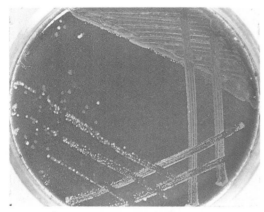

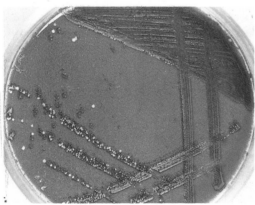

Fig. 42 Oxidase reaction. The heated blood agar medium had been inoculated with pus from the eyes of a two-day-old baby and incubated in an atmosphere of 5% CO_2 for 48 h at 37°C.
The white, pigmented colonies proved to be those of *Staph. epidermidis* (coagulase –ve); the lower photograph was taken 7 s after the oxidase reagent had been flooded over the plate. The purple-coloured colonies belong to the genus *Neisseria* and biochemical testing proved them to be *N. gonorrhoeae.*

hydrochloride) over the culture plate (Fig. 42). Neisseriae turn a rapidly deepening purple colour; it is essential to subculture such colonies within 3 min of applying the reagent if they are to survive for further study.

A more convenient method is to soak a piece of filter paper in the freshly prepared oxidase reagent and then rub into the impregnated filter paper a fleck of culture from a colony; thus, if the test is positive, the remainder of the culture is available for subculture and further study without exposure to the lethal reagent (Fig. 43).

Differentiation of the two pathogenic

Fig. 43 The filter paper is impregnated with oxidase reagent and a fleck of culture of neisserial growth is rubbed into the upper portion; a contrasting negative oxidase test is shown on the lower half

members of the genus from each other and from commensal species is readily performed by simple fermentation reactions; the carbohydrates under test must, however, be prepared in serum broth to satisfy the nutritional demands of the pathogenic members.

A more reliable and rapid method of performing fermentation reactions has been introduced; a dense suspension of overnight growth from a culture plate is made in a buffer-salt and aliquots of the suspension are added to 10% stock solution of the test sugars. The inoculated tubes are incubated at 37°C in a water bath and reactions can be read after 3 h (Table 2).

Table 2. Fermentative activities of some neisseriae

	Glucose	Glucose	Sucrose
N. meningitidis	⊥	⊥	–
N. gonorrhoeae	⊥	–	–
N. catarrhalis*	–	–	–
N. sicca	⊥	⊥	⊥

*Now belongs to the genus *Branhamella*

Serological characters

N. meningitidis can be assigned to one or other of nine serological groups by agglutination reactions with antisera prepared against cell-surface antigens; the most commonly encountered groups are A, B and C; group A strains account for the majority of cases of meningococcal meningitis in *epidemics* and are regarded as of higher pathogenicity than the other serogroups which are more usually found in the nasopharynx of healthy people.

N. gonorrhoeae are serologically heterogeneous although *fresh* isolates appear to comprise one serogroup.

Commensal members of the genus. All are oxidase positive and morphologically identical with meningococci and gonococci; all grow readily on ordinary nutrient agar and produce colonies in 18 h at 37°C, which serves to distinguish them from the two pathogenic species. Their colonies are thicker and more opaque than those of the pathogenic species. Some of the commensal members develop highly pigmented colonies.

NEISSERIAL INFECTIONS

As already stated only two members of the genus are pathogenic to man and both are obligate parasites with very poor powers of survival outside the human host.

N. meningitidis is the organism most commonly found in cases of acute pyogenic meningitis and although cases are usually sporadic, epidemic outbreaks are noted from time to time in Britain. In other countries, e.g. Nigeria, widespread epidemics are frequent during the dry season.

The meningococcus is rarely involved in other disease conditions but it is occasionally found in cases of conjunctivitis and endocarditis; similarly it can produce a septicaemia without any meningeal involvement and meningococcal septicaemia in an acute or chronic form must be remembered as a possible cause when cases of pyrexia of uncertain origin (PUO) are being investigated.

In the last few years effective vaccines have become available against group A and group C strains of *N. meningitidis* although, as yet, no satisfactory vaccine is available to protect against group B strains.

N. gonorrhoeae reflects its feeble viability outside the host by being almost always associated with adult cases of gonorrhoea which is transmitted by sexual intercourse; in males, acute gonorrhoea usually produces only an acute purulent urethritis but the organisms may spread to the prostate and epididymis. Adult female cases also suffer urethritis and the cervix is also infected and occasionally the infection extends to the endometrium and fallopian tubes.

Note that in the female a significant proportion (approximately 30%) of cases are symptomless and a majority suffer only trivial symptoms and do not even suspect infection.

Recent reports, particularly from Scandinavia, reveal that the incidence of pharyngeal gonococcal infection is increasing.

A bacteraemic phase may occur in both sexes with resultant gonococcal arthritis or rarely endocarditis.

Gonococcal infection of the infant's eyes occurs during childbirth if the mother is suffering from sexually-acquired infection, but gonococcal ophthalmia neonatorum is very much less common nowadays because of improved antenatal care and the detection and treatment of gonorrhoea when it occurs in the pregnant woman.

Older children, not yet capable of performing their own toilet and living in hospitals or other institutions may suffer infection of the eyes and, in girls, a vulvo-vaginitis; here the transmission is from an adult attendant suffering from gonorrhoea and who with slovenly habits of personal hygiene passes on the infection by means of damp sponges and towels.

N. gonorrhoeae strains capable of producing β-lactamase were first detected in 1976 and appeared to emanate from the Far East; the pandemicity of gonococcal infections saw the rapid spread to other countries and in Britain the incidence has increased from 15 isolates in 1977 to 211 in 1980 and 444 in 1981. No doubt these 'absolute' figures reflect a higher incidence, since not all β-lactamase producing strains are reported and many may not even be detected. The earlier importation of strains by individuals infected abroad has now given rise to an endemic situation within the UK. The therapeutic implications are obvious.

For the record it should be noted that β-lactamase producing strains of *N. gonorrhoeae* existed before 1976 since examination of freeze-dried cultures which were isolated in the 1940s has revealed a proportion of such strains.

Mycobacteria

9

This genus comprises various types of tubercle bacilli — two of which, the human (*Mycobacterium tuberculosis*) and bovine (*Myco. bovis*) types, are important pathogens of man — the leprosy bacillus, and many commensal and saprophytic species.

MYCOBACTERIUM TUBERCULOSIS

Microscopy
Size variable but approximately 3 μm × 0.3 μm, straight or slightly curved rods with rounded or slightly expanded ends. *Acid- and alcohol-fast* (Ziehl-Neelsen stain). Non-motile, non-capsulate, non-sporing (Fig. 44).

Fig. 44 Film of concentrated specimen of sputum stained by the Ziehl-Neelsen method. Acid- and alcohol-fast bacilli, in large numbers, contrast with the methylene blue background. *N.B.* Saprophytic members of the genus are only acid-fast.
× 800

Cultural appearances

Strictly aerobic. Pathogenic members grow *very slowly* even on rich media such as Lowenstein-Jensen (L–J) egg medium, and macroscopic growth rarely appears until at least 2–4 *weeks* incubation at 37°C (Fig. 45). The characters of *Myco. tuberculosis* on L–J medium can be summarised as 'rough, tough and buff' — i.e. rough = a dry and irregular surface; tough = hard and difficult to emulsify; and buff in colour, compared with the smooth, soft, readily suspended white growth of *Myco. bovis*.

Biochemical reactions

These have little relevance outside reference laboratories and for all routine practical purposes differentiation of *Myco. tuberculosis* from *Myco. bovis* and each of these from other potentially pathogenic types of mycobacteria is readily undertaken by simple criteria, e.g. speed of growth, optimum temperature for *in vitro* cultivation, etc.

Fig. 45 The macroscopic appearance of the human type of tubercle bacillus *(M. tuberculosis)* on Lowenstein-Jensen's medium, grown at 37°C for 6 weeks. The rough nature of the growth and characteristic colour are apparent

Serological characters

No valid tests are available for laboratory differentiation of cultures. Tuberculins derived from the cell wall are used to determine whether an individual has previously suffered from tuberculosis (or has been immunised with BCG); nowadays a purified protein derivative (PPD) of tuberculin is used. PPD is introduced into the skin of the patient's forearm and induration will occur at the site if the patient has experienced infection in the past, or is presently infected. Occasionally tuberculin testing gives a negative response if the patient is suffering acute disease, e.g. miliary tuberculosis, or is in the early stage of infection. Tuberculin testing is referred to by the technique used — Heaf test, Mantoux test or Tine test — and details of the tests and their interpretation can be obtained from medical textbooks.

Animal inoculation

If doubt exists whether an isolate from a patient is of the human or bovine type, the subcutaneous inoculation of a rabbit with the culture allows differentiation; bovine strains kill the animal within a few weeks and post-mortem examination reveals miliary spread. At most, human strains produce only a small lesion at the site of inoculation.

By contrast to the very occasional use of rabbits as outlined above, guinea-pigs are frequently used for diagnostic purposes; intramuscular injection, on the inner aspect of the thigh, with pathological material containing either human- or bovine-type bacilli leads to progressive disease. In communities where disease in the human population is under control, and many cases are 'spotted' early and excrete only few

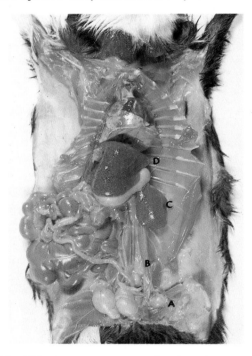

Fig. 46 Post-mortem appearance of a guinea-pig sacrificed one month after inoculation with a concentrated specimen of sputum which had not, at that time, shown any growth on Lowenstein-Jensen slopes. The lesions are as follows:

A = Superficial inguinal gland
B = Para-aortic glands
C = Slightly enlarged spleen with minute subcapsular tubercles
D = Liver with subcapsular tubercles

tubercle bacilli, guinea-pig inoculation is more sensitive than *in vitro* cultivation.

It is customary to inoculate a pair of guinea-pigs with material from the same individual; this covers the eventuality that one of the pigs may die from intercurrent infection or other disease. The animals are examined daily for evidence of local swelling or general infection; one of the pair should in any case be killed after four weeks and examined for local lesions, involvement of the deep inguinal and para-aortic lymph glands — rarely at this stage is the spleen involved either by enlargement or showing sub-capsular tubercular nodules (Fig. 46). If the first animal shows no evidence of disease its partner should be sacrificed after a further 4–8 weeks.

When any suspect lesion is noted smears must be made from it and stained by Ziehl-Neelsen's methods, since guinea-pigs suffer naturally from macroscopically similar lesions due to *Yersinia pseudotuberculosis* which is *not* acid-fast.

OTHER TUBERCLE BACILLI

Three other types of tubercle bacilli are recognised but only the avian variety has been incriminated (and very rarely) as a human pathogen. This type is morphologically identical with the mammalian types but it has a higher optimum temperature (42°C) and grows more rapidly on L–J medium, with production of light brown colonies, highly convex and glistening.

The vole variety is morphologically indistinguishable, non-pathogenic for man and very slow to grow even on L–J medium; particular interest surrounds this variety since it is as efficient in prophylaxis as the more widely used Bacille-Calmette-Guérin (BCG) vaccine which is an attenuated bovine strain.

A piscine or cold-blooded type of tubercle bacillus *(Myco. piscium)* is responsible for a granulomatous condition in fish, frogs, etc, but is non-pathogenic for man. Again these are morphologically similar to other types but have an optimum temperature of 22°C and resemble the avian type in cultural appearances.

ATYPICAL MYCOBACTERIA

These share the acid- and alcohol-fastness characteristic of *Myco. tuberculosis* and although they are occasionally isolated from pulmonary lesions they are most commonly associated with skin lesions or less often with cervical adenitis.

Classification of the atypical mycobacteria is not yet consolidated but they differ from *Myco. tuberculosis* and *Myco. bovis* in being non-pathogenic for guinea-pigs, in the colour of pigment produced and (frequently) in their speed of growth on L–J medium.

The normal habitat of most atypical mycobacteria is water or soil; *Mycobacterium marinum (balnei),* originally discovered in Sweden, causes skin ulcers and infection is often associated with tropical fish tanks.

Pigment production by atypical mycobacteria grown *in vitro* may be enhanced by exposure to light (photochromogens) or may occur under normal, dark incubation conditions (scotochromogens).

SAPROPHYTIC AND COMMENSAL MYCOBACTERIA

Saprophytic species are found on grasses and in milk and water; they are morphologically similar to tubercle bacilli but are not alcohol-fast and grow very rapidly and on ordinary media with production of various pigments. Non-pathogenic to experimental animals.

Particular interest surrounds *Mycobacterium smegmatis* (the smegma bacillus) which is commensal in sebaceous secretions particularly around the prepuce and labia; hence when urine specimens are being collected from suspect cases of genito-urinary tuberculosis careful toilet preparation reduces the number of smegma bacilli in the specimen.

Differentiation of this commensal from pathogenic tubercle bacilli can be made since it is not usually alcohol-fast and grows rapidly on ordinary media; similarly it is non-pathogenic for laboratory animals.

INFECTIONS CAUSED BY TUBERCLE BACILLI

Primary infection with *Myco. tuberculosis* usually results in a pulmonary lesion called the Ghon focus — this comprises a lesion which is frequently subpleural and also involvement of the hilar lymph nodes. In such instances the human-type bacillus is most commonly involved and infection is by inhalation.

Alternatively *Myco. bovis* ingested in raw milk may settle in the lymphoid tissue of the pharynx, and the cervical lymph nodes are also involved, or the site of entry may be the lymphoid tissue of the terminal ileum with lymphatic spread to the regional mesenteric lymph glands.

In all such instances the primary infection is frequently subclinical and healing usually occurs by fibrosis and calcification.

Infection with *Myco. bovis* is now uncommon in Britain and in other countries where preventive measures have been taken, i.e. the creation of dairy herds which are free from tuberculosis and/or the pasteurisation of milk has become commonplace.

Post-primary or adult tuberculosis is the infection which is seen clinically and may result either from a freshly acquired exogenous origin, particularly in *young* adults, i.e. under 40 years of age, or from a reactivation of an earlier healed lesion, as is more likely in older people.

The protective influence of immunisation with BCG is well established as a result of controlled trials carried out in Britain; conclusive evidence is available that BCG or alternatively vole vaccine gives a protection of at least 80% against natural infection and that, in particular, these vaccines are especially effective in preventing the more severe forms of human infection, i.e. meningitis and miliary tuberculosis.

The finding that a significant proportion of badgers in the South West of England suffer infection with *Myco. bovis* explains why cattle in that area have an incidence of bovine-type infection much higher than the national average and also underlines that we must constantly remind ourselves that the epidemiology of this and other infections is not static.

Indeed there is evidence of an increased incidence of pulmonary tuberculosis in certain groups of underprivileged, over-crowded populations in the UK, and although some contribution may have been made from immigrants there are areas where the increased incidence has no such association.

Finally, tuberculosis remains a hazard to certain occupational groups including laboratory workers, and although bacteriologists are more aware of the risks of laboratory-acquired infections, a recent report confirms that tubercle bacilli, e.g. in sputum, can survive the usual methods of heat fixation in making films for diagnostic purposes. Therefore such films should be stained immediately after preparation and if a delay is unavoidable, e.g. in countries where films may be made at local clinics and then be transported over long distances to a laboratory for staining and examination, the fixed smears should be rendered safe by immersion in glutaraldehyde.

LEPROSY

The causal organism is *Mycobacterium leprae*, similar in appearance to tubercle bacilli BUT less acid-fast and tolerates only 5% H_2SO_4 in attempted decolorisation in the Ziehl-Neelsen method. It is *not* alcohol-fast.

At present, identification is solely on microscopic grounds since the bacillus has not yet been cultivated; recently, inoculation of 9-banded armadillos with lepromatous material from patients has allowed an animal model to be developed for research purposes and this should allow rapid progress in the study of *Myco. leprae.*

For diagnostic purposes films are made from skin snips or nasal scrapings from affected patients, or sections of granulomatous lesions can be suitably stained; in such preparations, densely packed bundles of acid-fast bacilli can be seen lying free and also intracellularly.

Leprosy probably has a very low infectivity since although doctors and nurses often work for many years in leprosaria, infection of such dedicated individuals is exceptional. BCG vaccination is thought to give some degree of protection against leprosy but recent animal trials suggest that the use of a vaccine prepared from *Myco. leprae,* harvested from infected armadillos, will be more effective. There is clinical evidence that *Myco. leprae* has developed resistance to dapsone, the most useful therapeutic agent, and this is not without precedence but should stimulate even more the production of an efficient vaccine.

Corynebacteria

10

Classically only one member of the genus, *Corynebacterium diphtheriae,* is pathogenic to man; several species form part of the normal flora of the respiratory tract and other mucous membranes. Some species are pathogenic in animals and one of these, *C. ulcerans,* a pathogen of cattle, is occasionally associated with ulcerative throat lesions in the human host.

CORYNEBACTERIUM DIPHTHERIAE

Microscopy

Size very variable, essentially rod-shaped but often showing irregular expansion at one end — 'club-shaped'. Cells frequently lie in small clusters at acute angles to each other — Chinese-letter arrangement. Gram-positive but readily decolorised, non-motile, non-capsulate, non-sporing (Fig. 47).

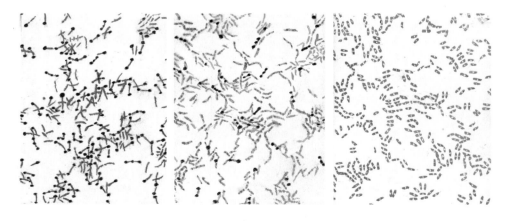

Fig. 47 Films, stained by Albert's method, of pure growths of diphtheria bacilli on the left, *C. xerosis* in the centre and *C. hofmannii.* The morphological similarity of *C. diphtheriae* and *C. xerosis* demands that they must be differentiated by further tests; *C. hofmannii* (on the right) shows its characteristic appearance; the absence of volutin granules and the clear unstained central bar are constant features of this organism
$\times$ 1000

Volutin granules can be demonstrated by Albert's staining method in film preparations from *luxuriant* media.

Cultural appearances

On Loeffler's serum medium, colonies are small (1–2 mm.), circular, grey and with a regular edge after 12–18 h incubation at 37°C; continued incubation gives colonies of larger diameter, with a crenated edge. Volutin granules are abundant in films from Loeffler's medium.

On blood tellurite media the bacilli give grey or black colonies since they reduce the tellurite within the bacterial cell; it is possible to recognise three colony types, *gravis*, *intermedius* and *mitis* on such media (Fig. 48).

Volutin granules are very scanty or absent in films made from tellurite media.

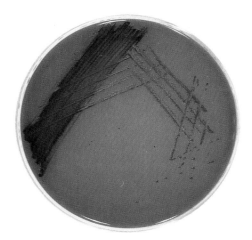

Fig. 48 Growth of *C. diphtheriae* on tellurite medium after incubation at 37°C for 38 h; the black nature of the colonies is typical

Biochemical activities

The reactions of the three colonial types of *C. diphtheriae* are contrasted with those of common commensal species in Table 3. It should be noted that only *gravis* strains ferment starch.

Tube haemolysis tests confirm the identity of the three biotypes of *C. diphtheriae; mitis* strains lyse ox and rabbit cells, *intermedius* are inactive against either blood and *gravis* strains lyse only rabbit cells.

Serological characters

The three bio-types are serologically distinct and each can be subdivided into several serological types by agglutination tests. Serotyping studies have been essentially academic but should prove of considerable epidemiological value.

Virulence is equated with the production of a powerful exotoxin and it is essential to demonstrate toxin production in a culture obtained from a throat swab before making a final report to the clinician. Non-toxigenic strains (= avirulent) are rare in *gravis* (1%), but more common in *intermedius* (7%) and *mitis* (15%) strains.

Diphtheria exotoxin is phage-specified and only those bacilli that possess β-phage are capable of producing toxin; *in vitro,* toxin production is also heavily influenced by the concentration of iron in the culture medium. We are able to render toxin *harmless* by incubating it at 37°C in the presence of formalin and the *toxoid* thus produced is used as an active immunising agent to stimulate the production of protective antitoxin in the human host.

Table 3. Biochemical activities of some corynebacteria

	Glucose	Sucrose	Starch
C. diphtheriae			
var. *gravis*	⊥	–	⊥
var. *intermedius*	⊥	–	–
var. *mitis*	⊥	–	–
C. hofmannii	–	–	–
C. xerosis	⊥	⊥	–

Animal inoculation

Two biologically equivalent guinea-pigs are used; one of these (control animal) is given an intraperitoneal injection of 1000 units of diphtheria antitoxin 12 h before inoculation and both guinea-pigs have their abdomens depilated; 0.2 ml of a 12 h culture of the organism suspended in broth is then injected *intracutaneously* into each pig.

Virulent bacilli produce a patch of erythema (10–15 mm in diameter) with ensuing oedema and within two days necrosis and then eschar formation follow. The control animal shows no reaction; reaction in both test and control animals indicates the organism is not a diphtheria bacillus.

Toxin production can also be demonstrated in a gel-diffusion test named after Elek or Ochterlouny who separately and simultaneously described the technique; in experienced hands this method is as reliable as the guinea-pig method, equally rapid and cheaper (Fig. 49).

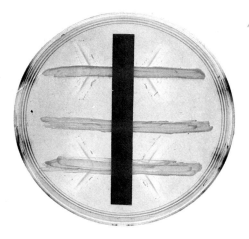

Fig. 49 Elek test for toxin production: The centrally-placed sheet of filter paper contains diphtheria antitoxin; positive (toxigenic) and negative (non-toxin producing) control strains were streaked across the upper and centre areas of growth. The lower strain under test shows lines of precipitation similar to those of the positive control and is a toxin-producing strain of *C. diphtheriae*

INFECTION CAUSED BY DIPHTHERIA BACILLI

Diphtheritic infections of wounds, etc, are occasionally seen but the usual site in which *C. diphtheriae* is found is in the faucial region, particularly the tonsils. There is little, if any, lymphatic spread and bacteraemia is unknown; other sites may be parasitised, e.g. the nose, larynx and ear, but regardless of site the clinical illness is caused essentially by the diffusible exotoxin which has a predilection for heart muscle and renal tissue. In addition the affinity of diphtheria exotoxin for certain nerves is reflected by post-diphtheritic paralysis of which palatal paralysis is the most common.

Diphtheria was endemic in Britain until the mid-1940s but was rapidly eliminated by the introduction of mass immunisation with diphtheria toxoid; occasional epidemics still occur but these are almost invariably traced to an imported source.

As with other infections which have been virtually eliminated from Britain by means of active immunisation, the disease is nowadays unknown to the younger generations who not unnaturally are less assiduous in having their children protected by active immunisation; hence a reducing proportion of the population are immune to diphtheria either as a result of natural infection or immunisation. The downward trend in acceptance of active immunisation was recently accelerated by emotive rejection of Triple Vaccine (diphtheria toxoid forms one of the three components) which was alleged to cause brain damage from the pertussis vaccine aliquot. Such a continuing reduction in herd immunity means that the population is increasingly susceptible to epidemic spread should a case occur. Associated with this danger is the fact that the vast majority of younger practitioners have never seen a case of diphtheria and not unnaturally may therefore not consider such a diagnosis, so that a missed case or one in which the diagnosis is delayed would not be isolated as quickly and would, therefore, act as a continuing source of infection to other individuals. Strenuous publicity is required to reaffirm the need for active immunisation and, for the doubters in our midst, we should also emphasise that active immunisation can be acquired against diphtheria (and tetanus) with a double toxoid preparation which does not include pertussis vaccine.

Bacilli

11

The genus *Bacillus* comprises species which are Gram-positive, aerobic and capable of forming endospores. *Bacillus anthracis,* the cause of anthrax in man and animals, was the first bacterium proven to be causally related to an infectious disease and until a few years ago was regarded as the only pathogenic member of the genus. Now *Bacillus cereus* is recognised as a much more common cause of human disease.

BACILLUS ANTHRACIS

Microscopy

A large organism (4–8 μm $\times$ 1–1.5 μm), rectangular although one or both ends may be

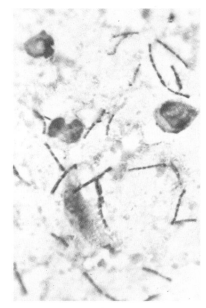

Fig. 51 M'Fadyean's reaction. Blood film from the peripheral blood of a cow dying of anthrax, stained with polychrome methylene blue. Short chains of anthrax bacilli are seen lying among irregular amorphous distintegrated capsular material. White blood cells are also present
$\times$ 1000

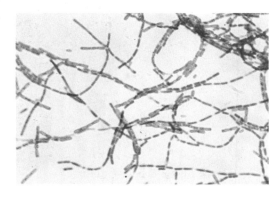

Fig. 50 Film of culture of *B. anthracis* stained by a modified Ziehl-Neelsen method, using 0.5% H_2SO_4 as the decolourising agent. Spores show as acid-fast material situated centrally within a bacillus and are oval in shape and non-projecting
$\times$ 800

concave; tends to occur in chains especially on *in vitro* cultivation. Gram-positive, non-motile, capsulate *in vivo,* spores are central, ovoid and non-projecting and are formed only *in vitro* (Fig. 50).

M'Fadyean's reaction (Fig. 51). Diagnosis of anthrax in animals is usually undertaken by microscopic examination of blood films. The application of polychrome methylene blue for 10–20 sec to a thick blood smear previously fixed by 1:1000 mercuric chloride for 5 min reveals large blue bacilli lying in a pool of disintegrated heliotrope capsular material.

Cultural appearances

Not in the least fastidious in regard to media, atmosphere or temperature requirements, although aerobic conditions are essential for sporulation and likewise the optimum temperature for spore formation is 25–28°C. Colonies on agar or blood agar are circular, have a 'medusa-head' appearance — a colony is one continuous thread of bacilli and has a ground-glass appearance, size 3 mm in diameter after 18–24 h at 37°C (Fig. 52) Only slightly haemolytic on blood agar.

Fig. 52 Growth of *B. anthracis* on blood agar after incubation at 37°C for 18 h; the ground-glass irregular colonies are characteristic. The irregularly contoured peripheral growth is typical

Biochemical reactions

These are rarely employed since the microscopic and cultural appearances of bacilli obtained from an infected animal or human are so characteristic.

Serological characters

The capsules of *B. anthracis* are *not* polysaccharide in composition as are those of many other bacteria, but consist of a polypeptide of D-glutamic acid of high molecular weight which is strongly antigenic; however anti-capsular antibodies, unlike those of the pneumococci, are without protective effect. Protein and polysaccharide somatic antigens have also been demonstrated but appear to play no greater a role than capsule substance in the natural disease. A lethal toxin is the prime factor in causing death and can be modified for use as an active immunising agent. No typing of *B. anthracis* for epidemiologic purposes has yet evolved.

Animal inoculation

The identity of *B. anthracis* can be confirmed by inoculating a guinea-pig with exudate from the lesion or a culture from such exudate; death with septicaemia occurs within 48 h and recovery of typical organisms from the heart-blood and spleen is diagnostic. Post-mortem examination of such animals must be undertaken with elaborate precautions to prevent any contamination of the operator or atmosphere and all surfaces must be thoroughly cleaned; disposal of the post-mortem board and carcase is by incineration.

BACILLUS CEREUS

This species was regarded as non-pathogenic to man until its recognition as a cause of toxic type food poisoning in the late 1960s and this role has stimulated intense study of the organism itself.

B. cereus is very similar to *B. anthracis* in morphology but is *motile;* also it is non-capsulate so that M'Fadyean's reaction cannot be elicited. Isolates can be allocated to one or other of 20 or more flagellar serotypes but most interest centres round the two enterotoxins which cause food poisoning.

SAPROPHYTIC MEMBERS

There are many such and they are usually spoken of as the *Bacillus subtilis* group, the latter organism being the type species. Many of these saprophytes are *motile* and the spores may be central, terminal or sub-terminal; haemolysis on blood agar is common; they do not exhibit M'Fadyean's reaction and are non-pathogenic for laboratory animals when injected in doses comparable with those of *B. anthracis.*

ANTHRAX

Anthrax occurs naturally in nearly all animals, and herbivores are the most susceptible; man is infected by spores shed into the environment by sick and dead animals and from spores contaminating animal products, e.g. wool, hides, bone-meal fertiliser. Infection from another human case is extremely rare. There are obvious occupational risks, e.g. to the farmer and veterinarian.

The internal types of the human disease, i.e. pulmonary and gastro-intestinal anthrax, are now extremely rare in Britain, the latter type never having been common. Pulmonary anthrax (wool sorter's disease) has been controlled by treating raw wool before it is processed and also by the provision of local exhaust ventilation which removes wool fibres from the worker's environment.

The external type of human anthrax, i.e. cutaneous anthrax, was once a hazard to porters at docks who humped imported hides, but this risk has been virtually eliminated by restricting the importation of hides and other materials and diverting them to ports where they can be mechanically unloaded and processed chemically to remove or destroy spores before they are handled by man.

Again, man is protected by strict legislation regarding the safe disposal of any animal which has died from anthrax in Britain.

The commonest source of cutaneous anthrax in this country nowadays is garden fertiliser made from bones contaminated with spores; epidemics of cutaneous anthrax have occurred in various parts of Britain in recent years and have been traced to the handling of bone-meal fertiliser.

Spores enter the skin through abrasions and after 2–3 days a papule appears which enlarges and although there is little or no erythema, the surrounding tissues are disproportionately oedematous compared with the small initial lesion. The painless nature of the lesion coupled with a knowledge of the patient's occupation and/or horticultural zeal should allow differentiation from carbuncles and erysipelas.

Death from cutaneous anthrax has become unusual since penicillin and other antibiotics are effective therapeutic agents.

TOXIC FOOD POISONING

Strains of *B. cereus* produce predominantly only one or other of the two enterotoxins; if the patient is affected by toxin A the onset of symptoms, vomiting and abdominal pain, is rapid, 1–5 h after ingestion of the incriminated food. Alternatively if toxin B is involved diarrhoea predominates with abdominal pain and the onset is delayed for some 8–15 h.

The short incubation type of illness predominates in Britain and the organisms, parasitic on rice, are killed off when rice is parboiled; spores survive such cooking so that if the boiled rice is left for some hours at room temperature, germination of the spores allows the emergent vegetative bacilli to multiply and elaborate exotoxin which can tolerate brief exposure even to high temperatures. Thus there is an association with fried rice and Chinese restaurants, but it must be appreciated that *B. cereus* is saprophytic on many other natural grains and foodstuffs.

Preventive measures are obvious in that foods which may be contaminated with *B. cereus* should be eaten immediately after cooking and if this is not possible then the cooked food should be immediately refrigerated to prevent germination of the spores which have survived cooking.

Finally a review of older literature suggests that *B. cereus* has been incriminated in wound infections and it may be that its essentially saprophytic role still encourages bacteriologists to regard this, and allied saprophytic bacilli, as an incidental finding unrelated to wound infection. We may still have new lessons to learn.

Lactobacilli

12

Lactobacilli occur commensally in all higher forms of life and in the human occur in the gut and vagina where they have valuable protective functions; they also thrive saprophytically.

Microscopy

Large bacilli (1–5 μm $\times$ 1.0 μm) often showing short chain formation. Gram-positive, non-motile, non-capsulate and non-sporing but pleomorphism is common *in vitro*.

Cultural appearances

Lactobacilli grow best at about pH 5.8 and can survive at even lower pH; this aciduric property allows the use of selective media to isolate strains from specimens, e.g. faeces, which have a myriad of bacterial species. Incubation on such selective media under anaerobic conditions for 48 h or longer reveals small colonies, 0.5 mm in diameter, with a granular surface and irregular edge.

Biochemical activities

At least nine species can be identified on the basis of their various abilities to utilise a range of carbohydrates; however the classification of the lactobacilli is still being argued.

Serological characters

Although serological analyses are the subject of research (and debate!) no useful information is presently available to assist species identification for diagnostic purposes.

THE ROLE OF LACTOBACILLI

In both sexes lactobacilli colonise the alimentary tract within a day or two of birth and remain, in large numbers, throughout life although their numbers may be dramatically reduced temporarily following the administration of broad spectrum antibiotics, e.g. the tetracyclines. The suppression of the lactobacillary flora can be disadvantageous to the patient whose microflora may be so upset that monilia and/or staphylococci become dominant and he suffers superinfection with such species.

Lactobacilli also dominate the commensal vaginal flora, particularly between the menarche and the menopause, when the glycogen content of the vaginal epithelium is an ideal energy source for lactobacilli which thus produce lactic acid to give a high acid (pH 4.5) vaginal secretion; this environment is antagonistic to microorganisms other than aciduric bacteria and yeasts and explains the virtual absence of vaginal infections during the child-bearing years.

Recently lactobacilli have been tentatively incriminated in urinary tract infections but the debate on their significance is just beginning.

Dental caries

The aciduric nature of lactobacilli dictates their role in dental caries; the primary attack on the dental enamel is probably by salivary streptococci and lactobacilli then participate in decalcifying the dentine. Both species act by producing acids by fermentation of dietary carbohydrates, particularly sucrose, and certainly gnotobiotic rats fed on a cariogenic diet and then infected with aciduric species develop dental caries.

In addition there are occasional reports of species of lactobacilli being isolated from patients' specimens, e.g. from blood culture, where their role as pathogens can hardly be doubted, and such reports receive backing from work with experimentally infected animals.

Clostridia

13

These are anaerobic, Gram-positive, spore-forming bacilli and the majority of species in the genus *Clostridium* are saprophytic and found in decomposing organic matter, soil, etc; some are commensal in the intestinal tract of man and animals and a few species are pathogenic. Those causing diseases in man are *Clostridium tetani, Cl. perfringens (Cl. welchii), septicum* and *oedematiens* and *Cl. botulinum.*

The species *Cl. difficile* has attracted attention in the last few years.

CLOSTRIDIUM TETANI

This organism, the cause of tetanus in man and many animals, is found in soil, particularly in manured soil, and is excreted from the intestinal tract of most animals including man; hence the potential presence of tetanus spores in cat-gut derived from sheep's intestine. In human cases of tetanus, the organisms are localised at the point of entry and the disease is essentially due to the powerful neurotoxin.

Microscopy

Slender rods with round ends, 5 μm × 0.5 μm, Gram-positive, motile with peritrichous flagella, non-capsulate. Spores are spherical,

Fig. 53 Film of culture of *Cl. tetani* stained by a modified Ziehl-Neelsen method, using 0.5% H_2SO_4 as the decolourising agent to demonstrate spores
× 2000

terminal and project widely — drum-stick bacillus (Fig. 53).

Cultural appearances

Strictly anaerobic and rapidly killed by normal atmosphere. Wide temperature range: optimum 37°C. On solid media isolated colonies are rarely seen since the organism spreads as a diaphanous film which might escape brief examination; the edge of the spreading growth shows numerous projections (Fig. 54).

Grows readily in cooked-meat medium with only slight digestion.

Fig. 54 Blood agar plate viewed by oblique illumination. *Cl. tetani* had been inoculated in a well area on the right and after 24 h incubation under anaerobic conditions the diaphanous spreading film of growth is readily noted and at the left edge numerous irregular projections of growth typify this species

Biochemical activities

No sugars are fermented; indole is produced.

Serological characters

Ten serotypes are distinguished on the basis of flagellar antigens. In all types, an identical neurotoxin (tetanospasmin) is produced, but some otherwise typical strains are non-toxigenic.

Animal inoculation

This is the most reliable laboratory technique for diagnosis. A pair of mice is used for each test; one mouse is protected by subcutaneous injection of 750 units of tetanus antitoxin 2 h before inoculation. Both animals are inoculated intramuscularly in the right hind leg; the inoculum comprises 0.25 ml of supernatant from a 48 h cooked-meat culture of the organism. Within a few hours, evidence of tetanus develops in the unprotected mouse — the tail becomes stiff, the inoculated limb progressively paralysed; thereafter the paralysis becomes more generalised and tetanic convulsions can be elicited by slight stimuli.

CLOSTRIDIUM PERFRINGENS (CL. WELCHII)

This organism is the commonest of several members of the genus *Clostridium* associated with gas gangrene. The main source of *Cl. perfringens* is the excreta of animals, including man. It must be appreciated that, as in the case of tetanus, the diagnosis of gas gangrene is made on clinical grounds since the presence of *Cl. perfringens* in a wound is, in itself, of no significance; classically gas gangrene is a rapidly spreading infection with oedema, necrosis, gas production and tissue gangrene.

Microscopy

Relatively large bacilli, 5 μm × 1 μm, with square or rounded ends. Gram-positive, non-motile (all other *Clostridia* are motile) capsulate in animal tissues. Spores are oval, subterminal and non-projecting (Figs. 55 and 57).

Cultural appearances

Not as strictly anaerobic as *Cl. tetani;* grows very rapidly. Colonies on blood agar are approximately 3 mm in diameter after 18–24 h incubation, semi-translucent with an entire

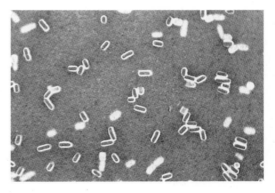

Fig. 55 Capsules of *Cl. perfringens.* This preparation results from mixing the bacteria in India ink and after drying the thin film in air it is fixed with methyl alcohol; subsequently Gram's methyl violet solution is applied and allowed to act for two minutes. Finally the preparation is washed in water, blotted and then dried over a flame
× 800

Fig. 56 Blood agar plate inoculated with *Cl. perfringens* and incubated anaerobically at 37°C for 18 h; colonies are entire, semitransparent and show haemolysis
× 800

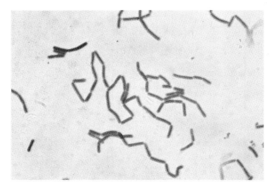

Fig. 57 Gram-stained film of *Cl. perfringens*. Spores, which would appear as unstained areas within the bacilli, are absent; this underlines the fact that sporulation in this species is not readily induced when grown on ordinary culture media
× 2750

Fig. 58 Stormy clot reaction. On the left is an uninoculated tube of litmus milk; that on the right has been inoculated with *Cl. perfringens* and incubated anaerobically for 24 h at 37°C. The milk has been clotted with acid formation, and gas production has broken up the clot. The yellowish fluid is the whey of the milk. The stormy clot is not as highly specific for *Cl. perfringens* as is sometimes believed since certain other members of the genus give similar reactions in litmus milk

edge (Fig. 56). Many strains show zones of β-haemolysis.

Biochemical activities

Ferments glucose, lactose, maltose and sucrose with much gas production. Indole is not produced. In litmus milk, rapid fermentation of the lactose occurs with the appearance of a 'stormy clot' reaction (Fig. 58)

Serological characters

Five types (A–E) may be differentiated according to the exotoxins produced; type A is that most commonly associated with gas gangrene, the other types are more commonly associated with animal diseases.

Nagler's reaction. All types of *Cl. perfringens* produce opalescence in egg-yolk media, due to production of lecithinase C (phospholipase) which causes a visible precipitate. Such opalescence is specifically inhibited by *Cl. perfringens* antitoxin; thus plates of the medium with one half of the surface coated with antitoxin and then inoculated on both halves allows identification of the organism (Fig. 59). Several other *Clostridia* give similar reactions on Nagler

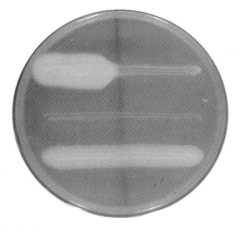

Fig. 59 Nagler's reaction. The egg yolk medium was smeared with *Cl. perfringens* Type A Antitoxin on its right half. The upper growth is that of *Cl. perfringens* and the opalescence of lecithinase activity (on the left) has been inhibited by the specific antitoxin; the central growth is of a non-lecithinase producing clostridium and the lower growth line is of a clostridial species whose opalescent growth has not been inhibited by the specific type-A antitoxin

plates and inhibition of opalescence can be demonstrated with their specific antisera.

Of the *Clostridia* considered in this chapter *Cl. botulinum* (all types) and *Cl. oedematiens* (except type C) produce opalescence but in the case of these other organisms the opalescence is not inhibited by specific antiserum except for that produced by *Cl. oedematiens* type A.

Animal inoculation

Virulence varies with different strains; with pathogenic strains the intramuscular injection of 1 ml of a fresh cooked-meat broth culture into a guinea-pig causes death in 1–2 days. Within a few hours of injecting a limb, marked oedema occurs and frequently crepitation can be detected; oedema and gas formation spread rapidly and the organisms are recoverable from the heart-blood within 12 h of inoculation. The administration of *Cl.*

perfringens antitoxin before injecting the culture fluid protects a control guinea-pig.

Of the several other *Clostridia* associated with gas gangrene, *Cl. oedematiens* and *Cl. septicum* occur most frequently.

CLOSTRIDIUM OEDEMATIENS

Microscopy

In stained films two features serve to differentiate *Cl. oedematiens* from *Cl. perfringens;* these are the projecting nature of the spores of the former and the fact that they usually stain Gram negatively. It is, however, motile and non-capsulate.

Cultural appearances

Strictly anaerobic and all strains tend to show spreading growth on solid media; of the four sero-types, A–D, all except those of type C are haemolytic on blood agar. Serotype B and D strains require the presence of cysteine and dithiothreitol for reliable surface growth to take place.

Fig. 60 Cooked meat medium. The centre tube was not inoculated while that on the right was inoculated with *Cl. perfringens* and the meat is not digested but slightly reddened (due to saccharolytic activity). The left-hand tube shows the gross proteolytic activity of the other clostridial species, e.g. *Cl. sporogenes*

Biochemical activities

Fermentative reactions are wide and may be variable although glucose and maltose are utilised by all types except type D which ferments only glucose and contrarily this is the only type which produces indole.

Serological characters

Types A, B and D produce exotoxins which are antigenic although cross-reactions with specific antitoxins can occur. Type C does not produce toxin in culture; type A strains can cause gas gangrene in man whereas type B and D strains predominantly affect animals.

Animal inoculation

Guinea-pig inoculation and protection tests can be performed in a manner similar to those for *Cl. perfringens* but cross reactions between the serological types may make interpretation difficult.

CLOSTRIDIUM SEPTICUM

Microscopy

Tend to be more slender than *Cl. perfringens* and citron-shaped organisms or filamentous forms occur. Spores are oval, subterminal and projecting.

Cultural appearances

Strictly anaerobic. Colonies are 3 mm in diameter after 48 h incubation, irregularly circular with a filamentous border. Growth tends to spread over the medium.

Biochemical activities

Glucose, maltose, lactose and salicin are fermented. Indole is not produced.

Serological characters

Four groups are recognised on the basis of somatic antigens and each group is further divided by the possession of various flagellar antigens.

Animal inoculation

Guinea-pigs are susceptible to inoculation and can be protected with antitoxin.

An almost identical organism, *Clostridium chauvoei,* is pathogenic only for animals. Biochemically it ferments sucrose and not salicin; serological relationships with *Cl. septicum* are recognised.

CLOSTRIDIUM BOTULINUM

This organism, essentially a saprophyte, is widely distributed in nature and quite frequently present in the intestinal tract of domestic animals. Botulism in man is an intoxication resulting from ingestion of foodstuffs containing toxin produced by the organism growing in the food.

Microscopy

Large, straight-sided rods some $4 \ \mu m \times 1 \ \mu m$ in size. Gram-positive, motile, non-capsulate. Spores are oval, central or subterminal in position and projecting — not unique in any morphological respect (Fig. 61).

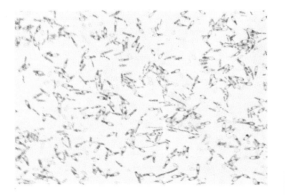

Fig. 61 Methylene blue stained preparation of *Cl. botulinum* in which the spores remain unstained within the vegetative cells. The spores are oval, subterminal and slightly projecting
× 2000

Cultural appearances

Strictly anaerobic — there is much variation in colonial morphology but colonies are often irregularly round with fimbriated edge, some 3 mm in diameter, translucent. Optimum temperature 35°C.

Biochemical activities

The six types, A–F, vary in their saccharolytic and proteolytic activities.

Serological characters

The six types produce neurotoxins which are antigenically distinct so that specific antitoxins can be used in guinea-pig tests to determine the type of the isolate.

Animal inoculation

Botulism can be demonstrated by intra-peritoneal injection of a fluid culture into a guinea-pig; a few hours after inoculation dyspnoea occurs followed by flaccid paralysis of the abdomen and ultimately generalised paralysis so that the animal becomes quite inert. As with similar tests for *Cl. tetani* paired animals are used, one protected with polyvalent botulinus antitoxin. Determination of the type of neurotoxin can be made by employing additional guinea-pigs each protected with type-specific antitoxin to the five types of neurotoxin.

Man is only affected by *Cl. botulinum* types A, B and E.

CLOSTRIDIUM DIFFICILE

This species was originally described in 1935 but only became the subject of intensive study in 1976; further prospective clinical and bacteriological investigations are necessary before its role in human infection can be clarified.

Microscopy

No unique characteristics to differentiate strains from other clostridia.

Cultural appearances

Strictly anaerobic. At present a selective medium, cefoxitin cycloserine fructose agar, is recommended for isolation of this species from faecal specimens.

Biochemical activities

These are shown in Table 4.

Serological characters

Most interest surrounds the production of a toxin which produces a cytopathic effect on monolayers of human embryo fibroblast cells; this activity can be neutralised with antitoxin produced by another species, *Cl. sordellii.*

CLOSTRIDIAL INFECTIONS

Gas gangrene

Although *Cl. perfringens* is the most frequently encountered species in gas gangrene other members of the genus, and

Table 4. Biochemical reactions of certain clostridia

	Glucose	Lactose	Maltose	Sucrose	Indole
Cl. tetani	−	−	−	−	+
Cl. perfringens	+	+	+	+	−
Cl. oedematiens	+	−	+(D)	−	−(D)
Cl. septicum	+	+	+	−	−
Cl. botulinum	+	−	+	−	−
Cl. difficile	+	−	⊥	⊥	−

D = *Cl. oedematiens* type D does not ferment maltose but does produce indole

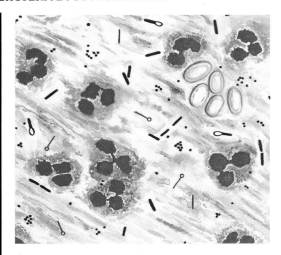

Fig. 62 This Gram-stained film of material from a wound shows pus cells (degenerate polymorphonuclear leucocytes), a few necrotic muscle fibres and clumps of Gram-positive cocci which on culture proved to be coagulase positive staphylococci.

Two species of *Clostridia* were also isolated: *Cl. tetani*, which in the film is represented by the slender Gram-positive rods and also rods with spherical, terminal and projecting spores which have stained Gram-negatively; such variation in reaction to Gram's staining method is well recognised. The stouter Gram-positive bacilli, some of which bear oval, sub-terminal, projecting spores, were finally identified as *Cl. oedematiens*

particularly *Cl. oedematiens* and *Cl. septicum,* are also incriminated in this infection; more than one species is frequently found in cases of gas gangrene (Fig. 62).

Infection results from the contamination of a wound by soil, dirty clothing, etc, but not all wounds from which such organisms are recovered develop gas gangrene and the diagnosis and initiation of treatment therefore must remain the responsibility of the clinician; however, endogenous infection is common in civilian practice because faecal carriage of *Cl. perfringens* is the rule.

On clinical grounds we can recognise three types of involvement of gas gangrene organisms; firstly a *simple contamination* of a wound as noted above with healing taking place without any evidence of infection; alternatively the organisms may multiply and spread along intermuscular septa but without invading uninjured, healthy muscle. The term

anaerobic cellulitis is used for this localised condition.

Finally the organisms may spread readily and with extremely invasive powers into healthy muscle which is destroyed by the powerful exotoxins; this is classical *gas gangrene* or clostridial myonecrosis.

Active immunisation against gas gangrene is not practicable because of the various species involved; polyvalent antiserum is available for prophylactic emergency use and for therapy. In an established case and if the causal organism has been identified several monovalent sera are available for therapeutic use but neither serotherapy nor antibiotic therapy can replace adequate surgical treatment.

Food-poisoning

Certain strains of *Cl. perfringens* can cause an infecto-toxic type of food-poisoning and such strains can be serotyped by agglutination testing with specific antibody-containing sera developed against the different polysaccharide capsular antigens.

Cl. perfringens food-poisoning is classically associated with pre-cooked meat dishes which are allowed to cool and then stored at temperatures allowing the causal organism to multiply (a stew is really a large cooked meat broth!) and elaborate enterotoxins; when such a dish is ultimately ingested the vegetative cells present undergo sporulation in the gut and *during sporulation* an enterotoxin is formed which produces symptoms of diarrhoea and abdominal cramps some 9–14 h later. Pyrexia is unusual, as is vomiting.

Tetanus

Tetanus, commonly referred to as lockjaw, follows the implantation of tetanus spores in a wound where conditions are suitable for the germination and the growth of the vegetative cells.

As already stated tetanus bacilli are commensal in the intestinal tract of many mammals including man and tetanus spores are thus ubiquitous and, in particular, they are found in large numbers in manured soil.

The clinical condition is caused by the powerful exotoxin, *tetanospasmin,* produced by vegetative cells in the injured area; the bacilli remain localised in the wound.

Tetanus neonatorum follows infection of the umbilical stump and this form of tetanus is most common in communities where cow dung and such like are used as umbilical dressings; it can, however, occur in more sophisticated communities if, as a result of inadequate sterilisation, tetanus spores survive in cord dressings or talc used to powder the umbilical area. Although tetanus occurs classically in the deep, dirty wound contaminated with soil and with necrotic tissues, many cases occur in some countries, e.g. Nigeria, where there is no obvious injury. It is considered that in these instances the spore is implanted in minor cracks or abrasions on the soles of the feet in those who walk bare-footed.

Similarly in Britain cases occur in which the trauma is minimal, e.g. a thorn prick or a simple superficial skin abrasion; such instances underline the the need for the much wider use of active immunisation with tetanus toxoid; such a practice would also reduce the use of tetanus antitoxin for emergency passive immunisation in the injured and thus lower the risks of hypersensitivity reactions, which are a recognised hazard of the administration of this and other antiserum preparations.

In any case, and even in individuals who have been actively immunised, adequate surgical treatment of all wounds remains the sheet anchor of tetanus prophylaxis.

Botulism

Fortunately this disease is rare in Britain; it results from the ingestion of foodstuffs, particularly preserved foods, in which *Cl. botulinum* has been growing and produced its hightly lethal exotoxin, which affects specifically the parasympathetic nervous system. Although the exotoxin is most potent, its clinical effects are not evident until 12–36 h after the incriminated food has been ingested and occasionally even 3 or 4 days may elapse before symptoms appear; in general, the shorter the incubation period the more severe the disease is and the higher the fatality; death is due to cardiac or respiratory failure.

Prevention of botulism demands that preserved foods must be adequately prepared and stored and home-canning in particular should be avoided; toxoid preparations are available for active immunisation but the very low incidence of the disease does not justify their use in man.

Infantile botulism has recently been described in the USA and a few cases have occurred in other countries; these are usually sporadic cases and in babies less than six months old; honey has been incriminated as the vehicle in some instances and the disease is usually milder than in adults.

Antibiotic-associated colitis

Diarrhoea or at least looseness of stools is a not uncommon side effect associated with antibiotic therapy, particularly with broad-spectrum antibiotics; a more specific association of colitis apparently caused by *Cl. difficile,* and often after treatment with clindamycin, has attracted much attention in recent years. This association is by no means absolute since clindamycin, in full dosage, can be given for up to six weeks, e.g. in cases of acute staphylococcal endocarditis, without any signs or symptoms of bowel upset.

Furthermore, *Cl. difficile* is now implicated in non-antibiotic-associated colitis and some such outbreaks suggest that patient to patient spread occurs; the epidemiology of *Cl. difficile* infections has not yet been clarified fully and it may be that this organism has a wider range of enteropathogenic effects than is presently obvious since non-cytotoxic strains have been isolated from cases of diarrhoea and treatment with vancomycin, in severly ill patients, has led to cure and coincident elimination of *Cl. difficile* from the faeces.

Enterobacteria

14

The family Enterobacteriaceae comprises numerous inter-related genera all of which are microscopically indistinguishable, 3–5 μm $\times$ 0.5 μm, relatively straight rods with rounded ends, Gram-negative, variously motile, some are capsulate, none possesses spores. All members of the family ferment glucose, with or without gas production. Their natural habitat is the intestinal tract of man and animals; some, e.g. *Escherichia coli*, are part of the normal flora whilst others, e.g. shigellae and salmonellae, are pathogenic for man.

Not all the genera of this family are considered in this volume.

ESCHERICHIA COLI

Microscopy

Strains have the general characteristics mentioned above; those which are motile possess peritrichous flagella.

Cultural appearances

Grows readily on all ordinary media and is aerobic and facultatively anaerobic. Colonies are 2–4 mm in diameter after 18–24 h at 37°C, opaque and convex with an entire edge.

On MacConkey's medium colonies are rose-pink on account of lactose fermentation (Fig. 63); they grow poorly if at all on deoxycholate citrate agar (DCA) with small dense pink colonies.

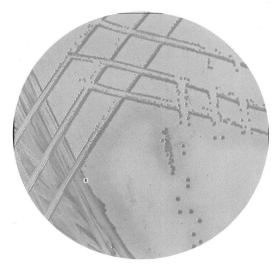

Fig. 63 MacConkey's plate after overnight incubation 37°C. The plate had been seeded with pus from a pilonidal sinus; characteristic lactose fermenting colonies of *Esch. coli* in profuse, pure culture

Biochemical activities

The day-to-day practical importance of these tests lies in differentiating *Esch. coli* strains, which are constantly present in human faeces, from the microscopically similar common pathogenic species of shigellae and salmonellae. More than 95% of *Esch. coli* strains ferment lactose promptly and with gas production so that confusion with *Shigella* strains rarely occurs. Virtually all strains produce indole so that this feature alone prevents wrong identification as a member of the genus *Salmonella* which never produces indole; the absence of urease activity separates *Esch. coli* from members of the genus *Proteus* at an elementary level. As in the identification of any species within the Enterobacteriaceae the final decision rests on serological differentiation.

Serological characters

This genus has been subjected to extensive antigenic analysis and more than 140 somatic (O) antigens have been identified and 49 flagellar (H) antigens are known. Routine serotyping is not undertaken except when strains are implicated in gastroenteritis; these enteropathogenic strains possess K antigens — present in capsules or microcapsules. Three types of K antigen, L, A and B, can be differentiated by their stability in various physical tests. Almost all enteropathogenic strains possess the B type of antigen.

Animal inoculation

This is not undertaken in diagnostic bacteriology but many strains are associated with natural disease in domestic animals and poultry.

BIOCHEMICAL INDENTIFICATION OF ENTEROBACTERIA

For many years the biochemical differentiation of the common bowel pathogens, shigellae and salmonellae, from

culturally similar non-pathogenic species has been made by inoculating colonies into a differential row of sugars. By using two composite media, *preliminary* identification of *Shigella* and *Salmonella* organisms can be effected with economies in time, labour and glassware (Fig. 64).

The pair of composite media shown in Figure 64A are uninoculated; tube 1 contains a mixture of glucose, mannitol and urea in a normal concentration of agar and is prepared as a slope with a deep butt. Tube 2 contains a mixture of sucrose and salicin in semi-solid

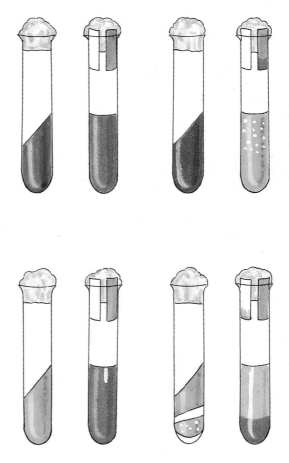

Fig. 64 Composite media for preliminary identification of some enterobacteria

agar over which lead acetate and Kovac reagent impregnated papers are suspended. Both media also contain relevant indicators.

The media are inoculated with a long, straight wire charged with a pure culture of the organism to be identified. Tube 1 is inoculated by both smearing the slant and then stabbing to the base of the butt; tube 2 is then inoculated by a single stab into its upper half inch after which the test papers are suspended above the medium and held by the cotton-wool stopper.

The tubes shown in Figure 64B reveal that the organism with which they were inoculated produces urease by the first tube showing a deep blue colour and also produces indole (yellow test paper) and utilises sucrose and/or salicin — such a pattern indicates that the culture was of the genus *Proteus*.

The reactions produced by *Shigella sonnei* and indole non-producing types of *Sh. flexneri* and *Sh. boydii* are shown in Figure 64C; glucose and mannitol have both been utilised in the first tube and without gas production; these species do not attack sucrose or salicin and are non-motile so that their growth in the second tube is restricted to the orginal inoculum track.

The first tube in Figure 64D shows gas production accompanying glucose and mannitol fermentation and in tube 2 motility is indicated by diffusion of the strain from the original inoculum line throughout the semi-solid medium; H_2S has been produced and is indicated by blackening of the lead-acetate paper. Indole has not been produced and no fermentation has occurred; the pattern of the reactions is biochemically consistent with a member of the genus *Salmonella*, other than *S. typhi*, which is anaerogenic.

SHIGELLAE

Members of this genus are the causative organisms of acute bacillary dysentery which is world-wide in distribution and endemic even in many highly developed countries including Britain. With the exception of certain simians man is the only host.

Microscopy

Identical with *Esch. coli*, but members of the genus *Shigella* are *never* motile or capsulate. Some strains of *Shigella flexneri* are fimbriate.

Cultural appearances

Similar to *Esch. coli* but give pale (colourless) colonies on MacConkey's and DCA medium since, with the exception of *Shigella sonnei*, they do not ferment lactose. Shigellae grow abundantly on DCA in comparison with *Esch. coli* (Fig. 65).

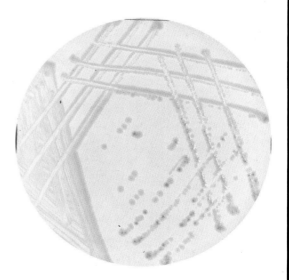

Fig. 65 MacConkey's plate sown with a mixture of *Esch. coli* (scanty, deep pink colonies) and *Sh. sonnei;* the plate has been incubated for 24 h at 37°C. For this reason the colonies of *Sh. sonnei* show a slight colour change from the pale, colourless state characteristic of shorter periods of incubation. *Sh. sonnei*, the commonest cause of bacillary dysentery in Britain, is a late lactose fermenter; all other members of the genus are lactose non-fermenting

Table 5. Groups within the genus Shigella

	Lactose	Mannitol	Indole	No. of serotypes
A. *Sh. dysenteriae*	–	–	V	10
B. *Sh. flexneri*	–	⊥	V	6
C. *Sh. boydii*	–	⊥	V	15
D. *Sh. sonnei*	(⊥)	⊥	–	1

Biochemical activities

Four groups are readily differentiated as in Table 5.

Group A strains are mannitol non-fermenting; members of Groups B and C must be differentiated by serological methods but it should be noted that *Sh. flexneri* serotype 6 strains are capable of biochemical subdivision and that certain of these are exceptional in producing gas (in small volumes). Group D strains are unique in fermenting lactose if incubated beyond 18–24 h and thus give pink colonies on MacConkey and DCA media.

Serological characters

Each of the groups is serologically distinctive; 10 serotypes are recognised within Group A and there is no significant intragroup relationship. Group B comprises six serotypes, all of which possess a common group antigen and each possesses a type-specific component.

Sh. flexneri types 1, 2 and 4 can be divided each into two subtypes on the presence or absence of minor group antigens. Group C contains 15 serotypes which are serologically distinct from Group B strains in lacking the *flexneri* group antigen. There are intra-group relationships so that for identification of the serotype of a *Sh. boydii* strain it is necessary to use absorbed antisera. Group D strains are a single serological entity. *Shigella dysenteriae* serotype 1 *(Sh. shigae)* produces a potent neurotoxin.

Animal inoculation

This has no value in the diagnostic identification of dysentery bacilli.

COLICINE TYPING

Colicines are naturally occurring antibiotics produced by many members of the family Enterobacteriaeceae; the colicine activity of a strain may be directed against other genera within the family although usually such activity is most commonly displayed against members of the same genus as that of the producing strain.

Use can be made of colicine production to characterise some organisms by noting various patterns of inhibition on a collection of passive or indicator strains (Fig. 66). This method is proving helpful in elucidating the spread of *Shigella sonnei;* this is the organism most commonly incriminated in cases of bacillary dysentery in Britain, an infection which has shown an almost unrelenting increase in incidence over the last thirty or more years.

Some of our inability to control this disease may have been associated with the fact that there is only one serological type of Sonne's bacillus so that all strains isolated from cases are serologically identical. We can now, by means of colicine typing, differentiate at least 17 stable types so that any particular type can be traced as it spreads through the population.

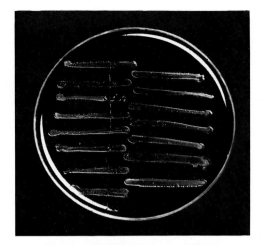

A

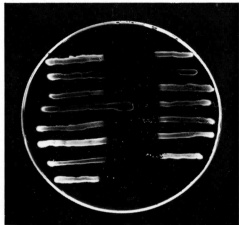

B

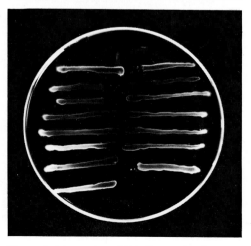

C

Fig. 66 Colicine typing of *Shigella sonnei* **A.** Type 7
B. Type 3 **C.** Type 2

TYPING TECHNIQUE

Production

The strain to be typed, i.e. the producer
strain, is diametrically streaked on tryptone
soya blood agar. The plate is then incubated
for 24 h at 35°C; the temperature of
incubation is important and should not be
allowed to rise or fall beyond 1°C from the
stated optimum, otherwise certain strains give
aberrant results.

Processing

After this period of preliminary incubation
the macroscopic growth of the producer strain
is removed with a microscope slide and the
microscopic remnants of growth are sterilised
by exposing the surface of the plate to
chloroform vapour for 10 min. Thereafter the
plate is exposed to air for a few minutes to get
rid of all traces of chloroform.

The 15 passive (indicator) strains are
streaked on to the plate at right angles to the
original growth line of the producer strain.
Eight strains are inoculated on the left-hand
and the balance of seven strains on the right-
hand side of the plate; this distribution serves
as a marker for the indicator strains.

Interpretation

A further period of incubation at 37°C (the
temperature is not now important) will allow
any colicines which have diffused into the
medium during growth of the producer strain
to exert their antibiotic activity on particular
indicator strains. The antibiotic activity of *Sh.
sonnei* type 7 is shown diagrammatically
(Fig. 67) and it will be noted that all the
indicator strains have grown with the
exception of strain number 3 which has been
fairly widely inhibited.

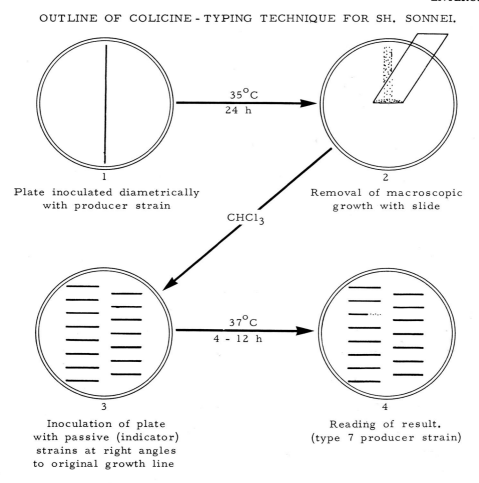

OUTLINE OF COLICINE - TYPING TECHNIQUE FOR SH. SONNEI.

1
Plate inoculated diametrically
with producer strain

35°C
24 h

2
Removal of macroscopic
growth with slide

CHCl₃

3
Inoculation of plate
with passive (indicator)
strains at right angles
to original growth line

37°C
4 - 12 h

4
Reading of result.
(type 7 producer strain)

Fig. 67 Outline of colicine typing technique for Sh. sonnei **1.** Plate inoculated diametrically with producer strain **2.** Removal of macroscopic growth with slide **3..** Inoculation of plate with passive (indicator) strains at right angles to original growth lines **4.** Reading of result (type 7 producer strain)

SALMONELLAE

There are more than 1500 serotypes within the genus *Salmonella;* four of these, *Salmonella typhi, S. paratyphi A, S. paratyphi B* and *S. paratyphi C* give rise to the enteric fevers and are parasitic only in the human intestine. The other serotypes occur widely as parasites of mammals and birds as well as of man, and the human infection associated with them is in the nature of acute gastroenteritis or bacterial food-poisoning.

Microscopy

Identical with *Esch. coli* but with the exception of *S. pullorum* and *S. gallinarum,* all types are motile and possess peritrichous flagella. Some species are fimbriate.

Cultural appearances

Aerobic and facultatively anaerobic; have a wide temperature range and like all enterobacteria grow readily on all ordinary media. Colonies are 2–4 mm in diameter after 18–24 h incubation at 37°C and on agar appear as greyish-white dome-shaped discs with an entire edge. On MacConkey's and DCA media colonies are similar in size and shape to those on agar and are pale since lactose is not fermented.

Biochemical activities

There is a profusion of biochemical tests but for normal diagnostic purposes it is sufficient to determine that the organism ferments glucose, dulcitol and mannitol with gas production (N.B. — *S. typhi* is anaerogenic), does not ferment lactose or sucrose and does not produce indole. On this evidence and the demonstration of motility further identification is by serological analysis.

However for epidemiological purposes biotyping of some species is valuable and indeed has been the accepted method of epidemiological typing for some species, eg. *S. enteritidis,* for more than 40 years. More recently a scheme for biotyping *S. typhimurium,* the species accounting for the majority of outbreaks of salmonella food poisoning, has been introduced; thus on the basis of five primary fermentation tests (using D-xylose in Bitter's medium, *meso*-inositol in peptone water, L-rhamnose in peptone water and performing a *d*-tartrate turbidity test and *m*-tartrate plate test), 32 primary biotypes of

S. typhimurium are recognised. Subtypes within these primary biotypes are distinguished by performing ten additional tests so that at least 144 full biotypes can be recognised.

Serological characters

All motile salmonellae possess two main antigens. The O or somatic antigens are Group antigens and it will be noted in Table 6 that identification of these with group-specific antisera allows the unknown organism to be allocated to one or other of the Groups. The H or flagellar antigens of a single species may occur in either or both of two phases; phase 1 antigen being shared by only a few other species whereas phase 2 is shared by many; these phases are referred to respectively as specific and non-specific phases. A few salmonellae possess a third antigen occurring on the surface and designated the Vi antigen; when present it masks agglutination with O antisera.

Bacteriophage typing. Certain salmonella serotypes can be subdivided into phage-types for epidemiologic purposes.

Animal inoculation

Species vary in their virulence for various laboratory animals; inoculation is of no importance in diagnostic laboratory practice.

Table 6. Salmonellae: some representatives of the genus (Kauffman-White schema)

Group	Serotype	Somatic antigens	Flagellar antigens Phase 1	Phase 2
A	*S. paratyphi A*	1, 2, 12	a	–
	S. kiel	1, 2, 12	g, p	–
B	*S. paratyphi B*	1, 4, 5, 12	b	1, 2
	S. typhimurium	1, 4, 5, 12	i	1, 2
C_1	*S. paratyphi C*	6, 7 Vi	c	1, 5
	S. thompson	6, 7	k	1, 5
C_2	*S. newport*	6, 8	e, h	1, 2
	S. bovis-morbificans	6, 8	r	1, 5
D	*S. typhi*	9, 12, Vi	d	–
	S. enteritidis	1, 9, 12	g. m	–

PROTEUS

Members of this genus are widely distributed in nature and are also found in the faeces of animals and man. They occur as pathogens in wounds, bed-sores and urinary tract infections; infection may be exogenous or endogenous in origin.

Microscopy

Similar to other enterobacteria but very pleomorphic. Motile.

Cultural appearances

Not in the least fastidious in regard to temperature, atmosphere or nutritional requirements. On nutrient agar these organisms rarely grow as isolated colonies but swarm in successive waves over the surface; swarming is inhibited on MacConkey's medium or by increasing the agar content of blood agar to 4% (Fig. 68).

Fig. 68 Blood agar plate which had been inoculated at one point with *Proteus vulgaris*. After 18 h incubation at 37°C four successive waves of growth can be seen; swarming of *Proteus* species delays the isolation of other organisms in mixed cultures but this characteristic can be inhibited by incorporating various substances in the medium, e.g., sodium azide (1 in 5000), chloral hydrate (1 in 500) or by increasing the agar content. Similarly the inhibition of swarming on MacConkey's medium is valuable in allowing separation of other species, provided that these grow on this selective medium

Table 7. Biochemical types of Proteus

	Mannitol	Maltose	Indole
Pr. vulgaris	−	+	+
Pr. mirabilis	−	−	−
Pr. morganii	−	−	+
Pr. rettgeri	⊥ or +	−	+

Biochemical activities

Four biochemical types can be recognised as shown in Table 7.

Lactose is not fermented so that on MacConkey or DCA media, pale colonies are noted (Fig. 69). Differentiation from the popular pale pathogens isolated on such media is effected by testing for urease activity. Proteus species decompose urea rapidly with the liberation of ammonia; shigellae and salmonellae do not produce urease.

Serological characters

Pr. vulgaris and *Pr. mirabilis* have been subjected to serological analysis and 119 serotypes can be recognised on the basis of O and H antigens, analogous to serotyping of salmonellae.

Certain types of proteus have antigens in common with the rickettsiae and this is the basis of the Weil-Felix reaction in the sero-diagnosis of the typhus fevers.

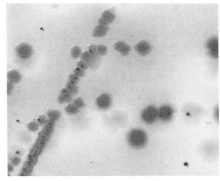

Fig. 69 This plate of MacConkey's medium was inoculated with a specimen of urine from a patient with paraplegia of several years' duration; the large and small lactose fermenting colonies are those respectively of *Esch. coli* and *Strept. faecalis*. The pale, lactose non-fermenting colonies are those of *Pr. mirabilis* on this medium
×2

KLEBSIELLAE

The majority of types within this genus are saprophytic, e.g. in water supplies. Other strains occur commensally in the intestinal tract in a minority of healthy people. The most common site in which *Klebsiella* strains fulfil a pathogenic role in man is the urinary tract.

Microscopy

Similar to other members of the Enterobacteriaceae but are non-motile. Capsulate both in the tissues and on *in vitro* cultivation.

Cultural appearances

Colonies are large, high-convex and mucoid on account of abundant production of extracellular slime; colonies tend to coalesce (Fig. 70). On MacConkey's medium the majority of strains give pink colonies due to lactose fermentation (Fig. 71).

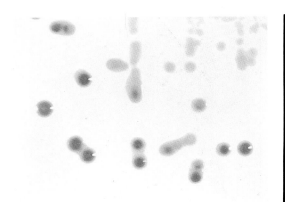

Fig. 71 This MacConkey plate shows, after overnight incubation at 37°C, the characteristic colonies of *Klebsiella* species; they are lightly pink in colour (due to lactose fermentation) and mucoid due to the production of extracellular slime. This latter feature also dictates the coalescence of adjacent colonies

Biochemical activities

Strain variation is so great that it defies summary treatment — in any case their characteristic colonies allow easy differentiation from other enterobacteria. The majority of strains are urease-producers but are much slower and less intense in this regard than proteus strains; their lack of motility and non-spreading growth on ordinary media further differentiate them from proteus.

Biotyping schemes are available which allow epidemiological investigation of outbreaks of klebsiella infection without recourse to the more expensive serological typing methods.

Serological characters

In addition to somatic antigens, strains also possess K (capsular) and M (mucoid) antigens; in any one strain the K and M antigens are identical. The presence of K and M antigens masks the O antigen so that capsular antisera are used to characterise strains. Thus the genus is divided into 80 defined serotypes by means of capsule-swelling reactions similar to the technique employed in typing pneumococci (Fig. 72).

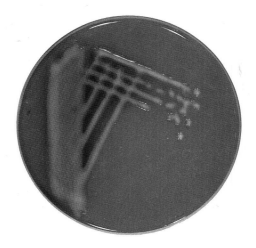

Fig. 70 Typical appearance of *Klebsiella* species on blood agar after 18 h incubation at 37°C. The mucoid nature of the colonies should be noted

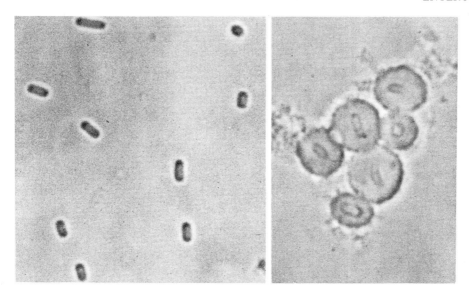

Fig. 72 Typing of *Kl. aerogenes* by capsule swelling reaction. On the left is a wet film preparation of *Klebsiella aerogenes*, type 54, mixed with *heterologous* antiserum: the *large* capsules, which were seen in parallel India ink films, are invisible. On the right is a wet film of the same organism mixed with its *homologous* antiserum and not only are the capsules 'swollen' but the bacilli are agglutinated

$\times$ 3750

INFECTIONS CAUSED BY ENTEROBACTERIA

Whilst some species of the Enterobacteriaceae cause certain well-defined intestinal syndromes, others are entirely commensal in the gut but are associated with infection in other tracts and tissues.

Intestinal infections

In Britain intestinal infections caused by enterobacteria are bacillary dysentery, salmonella food poisoning, the enteric fevers and also gasto-enteritis caused by enteropathogenic *Esch. coli.*

Bacillary dysentery

Bacillary dysentery, shigellosis, occurs throughout the world and has increased in incidence in Britain in the last 30–40 years so that it is now a common endemic bacterial infection in this country. The severity of

bacillary dysentery is, in general, related to the different groups of shigellae and the more severe illnesses are usually caused by members of the *Sh. dysenteriae* group; the clinical picture produced by the serotypes of *Sh. flexneri* and *Sh. boydii* is less severe and cases of dysentery caused by *Sh. sonnei* are frequently so mild that the infection amounts to little more than a social inconvenience and the individual can continue at work.

The fact that many if not most cases of Sonne dysentery remain ambulant is very probably associated with the increasing incidence of the disease in Britain since such individuals have very much greater opportunities to spread the disease than those who are confined to their own homes.

Spread is essentially by the hand-to-mouth method; the case contaminates toilet fixtures, including wash basin taps and door handles, etc, and the organisms are thus transmitted to the hands and then to the mouths of other people; very occasionally Sonne dysentery may be water-borne when in rural areas a sewage outfall contaminates a river from

89

which water is drawn for culinary and other human uses and is drunk without being efficiently purified.

Prevention, therefore, depends essentially on education of the individual in methods of maintaining a high standard of personal hygiene, and since the infection is most common in school age groups there is also a need for the provision of *adequate* toilet facilities in schools; in particular, wash basins should be in the same cubicle as the toilet pedestal so that there is no need for communicating doors which must be handled before the user can wash his hands.

Salmonella food poisoning

This infection-type food poisoning is caused by members of the genus *Salmonella* other than the four serotypes causing the enteric fevers; however, *S. paratyphi B* is exceptional in causing either an enteric fever which is a systemic infection *or* food poisoning. *S. typhimurium* has headed the league table in salmonella food poisoning for many years.

As in bacillary dysentery, infection is by ingestion; however, unlike dysentery, human cases and carriers are *not* the sole sources of infection; domestic animals, e.g. cows, sheep and pigs also act as sources and these also may be cases of infection or carriers. Rodents are heavily parasitised with salmonellae and can contaminate foodstuffs.

Meat intended for human consumption may be infected with salmonellae 'on the hoof' or it may originally have been free of salmonellae and become contaminated by various means during its preparaton and serving, e.g. carcases can contaminate each other in the abattoir or during transportation to wholesale and then retail butcher's premises and there, clean meat can be contaminated via knives, choppers, etc, unless these are carefully cleaned between carcases. Similarly the human case or carrier involved in dressing the carcase can contaminate the meat and such human sources can also act when the carcase is being reduced to saleable fractions and in the ultimate stages of

preparation at home or in kitchens for the preparation of communal meals, e.g. school-meal kitchens, hotels, etc.

Cow's milk and products derived from it may be a source of infection; in particular, artificial cream is frequently incriminated in epidemics and this material is an excellent culture medium for many organisms including salmonellae. Thus foodstuffs incorporating contaminated artificial cream and stored at normal room temperature for hours or days before consumption will have a much higher population of salmonellae than if they had been eaten immediately after preparation.

Bird's carcases, especially those of hens, ducks and turkeys, can also act as a source of infection for man since they suffer from salmonella infection; and eggs may be the source of infection. Eggs may be contaminated from cloacal discharge during the cooling and drying period immediately after they are laid, or in the case of duck eggs infection may occur in the oviduct whilst the egg is being formed. Infection arising from eggs will involve only the individual consumer unless eggs are pooled in the preparation of a communal food, e.g. custards, when eggs which are free from infection may be contaminated if one infected egg is in the pooled material.

Obviously the prevention of salmonella food poisoning demands careful supervision of meat and other materials intended for human consumption at all levels of preparation 'from the hoof to the home'. Rodent-proofing of premises prevents a further source of contamination and food which has been prepared but is not immediately consumed must be adequately stored, preferably in a refrigerator, so that any salmonellae which may be present do not have an opportunity to multiply.

The health of food-handlers should be constantly supervised and they must be educated and trained in methods of handling food which minimise or eliminate the risk of contamination.

The enteric fevers

Typhoid fever was endemic in Britain until water supplies, in addition to being filtered, were ultimately chlorinated so that at the point of delivery for consumption there is a small concentration of residual chlorine. Imported cases still occur but spread from such index cases is virtually unknown; man is the only source, as a case or carrier, of *S. typhi*. On occasion the role of water supplies may be indirect, as in the Aberdeen epidemic in the late 1960s when the vehicle of infection was canned meat which had been contaminated originally from water containing *S. typhi* used in the canning process in the country of origin.

Paratyphoid fever in Britain is almost always caused by *S. paratyphi B;* cases caused by *S. paratyphi A* or *C* strains are rare and usually occur in persons infected abroad.

In communities with sophisticated water supplies and other environmental services and where cases of enteric fever are readily recognised and treated in isolation, carriers of enteric fever salmonellae are more likely to act as sources than are cases of infection. There are few foodstuffs which have not been incriminated as vehicles of infection and whilst some of these are imported in a contaminated state, e.g. desiccated coconut, other foods are contaminated by the hands of carriers employed in the food industry.

Therefore prevention depends on the institution and maintenance of safe water supplies and adequate methods of sewage disposal. People employed in water works, milk production — including those on dairy farms — and in the production of food stuffs should be screened to ensure that they are not enteric carriers.

Reasonable protection can be provided by actively immunising individuals with monovalent acetone-preserved typhoid vaccine; such immunity should be offered to people living in or visiting countries with inadequate water supplies or sewage disposal systems but is not recommended for those living in well-developed communities.

Gastroenteritis

Infection with enteropathogenic strains of *Esch. coli* usually affects children in the first year of life and epidemics occur in residential nurseries and other institutions; the causal role of such strains of *Esch. coli* was recognised only some 35 years ago and babies who are artificially fed are much more likely to suffer infection than those who are breast-fed. Sources of infection are cases or carriers and spread is facilitated by fomites, particularly milk feeds which are not sterilised after being prepared and bottled; such *terminal sterilisation* eliminates any danger of infection by this method.

In the last few years there has been a definite decline in the reported incidence of gastroenteritis caused by entero-pathogenic *Esch. coli*.

Some strains of *Esch. coli* produce enterotoxins with a cholera-like action on the intestinal mucosa; other strains, equally associated with gastroenteritis, do not appear to be capable of producing such enterotoxins and the pathogenesis of such infections is still not unravelled.

Non-intestinal infections

Urinary tract infections. *Esch. coli* and other enterobacteria commensal in the gut are a common cause of urinary tract infections, and usually these are *endogenous* in origin; of particular significance is *silent bacteriuria of pregnancy*. The silent nature of such infections, i.e. without signs or symptoms of infection, demands that routine screening for such an infection is a mandatory part of antenatal care so that the 6–8% of women who are found to be infected can be treated to eliminate the infection. If such silent bacteriuria escapes detection a proportion of such infections continue and may result in ascending infection over the course of months or years and result in chronic pyelonephritis.

Urinary tract infections with enterobacteria may be acquired *exogenously* when the organisms are introduced as a result of instrumentation required for diagnostic or therapeutic purposes; instruments may not have been properly sterilised before use or the pre-operative cleansing of the patient may have been inadequate so that *Esch. coli* or other enterobacteria present in the ano-genital region are carried into the bladder by the catheter or cystoscope.

Catheterisation to acquire a specimen of urine for bacteriological examination should be *banned,* since such a procedure can cause infection!

Wound infections

The enterobacteria have assumed increasing significance in infection of surgical and other wounds, burns and bed sores, and epidemics of such infections in hospitals are now at least as common as those caused by *Staph. pyogenes.* Digital transmission from case to case is easily obtained unless personnel are scrupulous in their hand washing; inanimate objects in the hospital environment also play an important intermediary part in transmission of such infections.

Finally enterobacteria, particularly *Esch. coli,* have a significant role in neonatal meningitis, which is usually hospital-acquired and often iatrogenic.

Rare but spectacular cases of lobar pneumonia are caused by klebsiellae of serotypes 1 and 2; the serotypes are given the specific name of *Klebsiella pneumoniae* (Friedlander's bacillus) and such infections carry a high mortality rate.

Pseudomonads

15

Members of the genus *Pseudomonas* are primarily saprophytic and ubiquitous and are found in soil, water, on plant-life and elsewhere in man's environment. Although some species, particularly *Pseudomonas pyocyanea (Ps. aeruginosa),* have attained prominence in recent years as a cause of infection, particularly in the compromised patient, it is worth noting that at the beginning of this century a German clinical bacteriologist stated that *Ps. pyocyanea* was the greatest single threat to life in infancy!

PSEUDOMONAS PYOCYANEA

Microscopy

Relatively straight, Gram-negative rods, motile by virtue of polar flagella, non-capsulate, non-sporing.

Cultural appearances

Strict aerobe. Wide temperature range. Growth occurs on all ordinary media and on nutrient agar colonies are 2–4 mm in diameter, convex and with an entire edge although effuse growth is not uncommon. Two striking characteristics are the sweet musty odour and the green coloration which diffuses into the medium; the green coloration is due to the production of the pigment pyocyanin (Fig. 73). A minority of strains produce other pigments, e.g. red, and some strains do not produce any pigments. Certain strains are heavily mucoid.

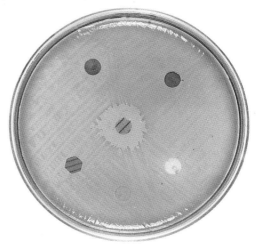

Fig. 73 Nutrient agar sensitivity test plate sown with a culture of *Ps. pyocyanea;* the natural colour of the medium (light straw) has been altered by pyocyanin pigment produced by the organism. Characteristically this strain shows resistance to many antibiotics, and indeed, in this case, is sensitive only to colomycin sulphate
$\times \frac{2}{3}$

Biochemical activities

Of the sugars commonly employed in diagnostic laboratories only glucose is utilised and without gas production.

All strains produce oxidase rapidly so that non-pigment producing variants can be readily identified; few other commonly occurring Gram-negative bacilli are responsive to the oxidase reagent and these latter react more slowly (2 min or more) compared with *Ps. pyocyanea* which responds within 30 s.

Serological characteristics

Strains possess somatic and flagellar antigens but no internationally agreed serotyping scheme has yet evolved; vaccines have been developed against *Ps. pyocyanea* which are giving encouraging results in field trials.

Pyocine typing

A method of typing strains for epidemiological tracing purposes has been developed and is now used in many countries; the technique is analogous to that of colicine typing of *Sh. sonnei* except that incubation of the strain to be typed, i.e. the producer strain, must be at 32°C for 14–18 h otherwise aberrant results will be given by certain strains. Similarly the eight indicator or passive strains are different from those used in colicine typing and all are *Ps. pyocyanea;* at least 37 different types within the genus can be recognised by their various patterns of inhibition on the indicator strains.

The patterns of inhibition illustrated in Figure 74 are those of pyocine type 1 and type 3 strains, and these are the types most commonly found not only in Britain but in many other countries.

Subdivision of type 1 strains can be made by an identical technique using five extra indicator strains.

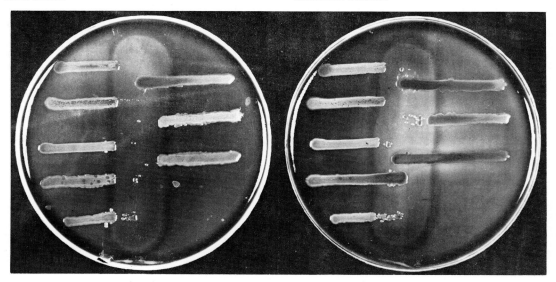

Fig. 74 These plates were originally inoculated in the vertical plane with strains isolated from patients in an Assisted Respiration Unit; the plates were then incubated at 32°C for 14 h and after the producer strain had been removed and the surface of the plate exposed to chloroform vapour the eight indicator strains were streaked on to the plate at right angles to the original growth line of the producer strains. The plates were then reincubated at 37°C for 18 h. Indicator strains 1 to 5 are on the left of each plate, indicator strain 1 being at the top, and similarly indicator strains 6 to 8 were streaked out on the right-hand side of each plate.

The type 1 producer strain has inhibited all of the indicator strains with the exception of strain no. 6; in the case of the type 3 producer strain indicator strains nos. 4, 6 and 8 have escaped inhibition

PSEUDOMONAS INFECTIONS

In addition to *Ps. pyocyanea* several other species, equally saprophytic and opportunistic, are incriminated in human infections.

The significance of *Ps. pyocyanea* in human infection is dependent primarily on its remarkable innate resistance to most antibiotics and its resistance to most disinfectants as well as its elementary physiological requirements for growth and survival; these features which combine to ensure its survival led a colleague to describe *Ps. pyocyanea* as 'the nearest thing to spontaneous generation known to man'. This apparently facetious comment is realistic since this non-sporing saprophyte is very difficult to eradicate from the hospital environment.

Ps. pyocyanea is particularly dangerous when it infects debilitated patients, e.g. those who have suffered multiple injuries or individuals receiving radiation therapy; in such people infection frequently extends from a local site to become septicaemic.

Infection of the urinary tract, which is almost invariably exogenous, is common in patients on continuous drainage, e.g. paraplegic patients, and is often refractory to antibiotic therapy; the ability of *Ps. pyocyanea* to survive and indeed to grow in many disinfectants and antiseptics explains the occasional epidemics in hospital practice, e.g. the use of multidose containers to administer eye-drops to patients who have undergone ophthalmic surgery has led to several recent epidemics of infection.

Another group of patients who are peculiarly susceptible to infection are those who require tracheostomy; it is essential that such individuals should be treated in strict isolation from each other and that respirators and other apparatus must be properly sterilised before being used for other patients.

Children suffering from cystic fibrosis are prone to pulmonary infections and *Ps. pyocyanea* is often incriminated; such strains are usually heavily mucoid when isolated but appear to be variants of normal non-mucoid strains as judged by identity of pyocine type of sequentially isolated strains which become mucoid as the infection progresses and the patient's clinical condition worsens.

To emphasise the resistance of this species it should be noted that bacteriologists use selective media, e.g. Cetrimide or Dettol agar plates for isolation of *Ps. pyocyanea* from environmental situations!

In recent years a new endeavour to combat infections with *Ps. pyocyanea* by active and passive immunisation is bearing fruit; two polyvalent vaccines, derived from *Ps. pyocyanea* cell wall extracts, have been assessed in certain groups of compromised patients, e.g. leukaemia and certain other malignancies where immunosuppressive therapy induces increased susceptibility to opportunist pathogens such as the pseudomonads.

Severely burned patients are also at risk from infection with *Ps. pyocyanea;* these and other compromised patients show a definite response to *active immunisation* with polyvalent vaccines in that infection rates and case fatality rates are reduced.

Similarly, significant protection from *Pseudomonas* infection can be achieved in burned patients who are *passively* immunised with *specific immunoglobulin* prepared from the plasma of healthy volunteers who have been actively immunised with polyvalent vaccines.

Bacteroides

16

Bacteroides species are not only a predominant part of the commensal flora of the lower intestinal tract, the mouth and the vagina, but play a significant role as pathogens after surgical procedures at these sites. Furthermore, they are efficient agents in causing infections in compromised patients.

Despite detailed analyses of large numbers of clinical isolates by several groups of authoritative research workers no agreed classification has yet been reached; indeed, detailed species differentiation remains not only time-consuming but costly and many routine laboratories are forced to report isolates as '*Bacteroides* species'.

But for the advent of the genus *Legionella*, members of the genus *Bacteroides* would have attracted more attention than any other group of bacteria in the last decade. A simplified classification allows at present the recognition of three groups within the genus, i.e. the fragilis group, the melaninogenicus-oralis group and the asaccharolytic group (Table 8).

BACTEROIDES FRAGILIS

Members of this group, which presently comprises nine species, are the most commonly encountered in human infection.

Microscopy

Rod-shaped, 2–3 μm × 0.4–0.8 μ, Gram-negative, non-motile, non-sporing with some strains being capsulate.

Cultural appearances

Strictly anaerobic; colonies are 1–2 mm in diameter with an entire edge and convex elevation; smooth, grey or greyish-white (Fig. 75).

Biochemical reactions

The most significant reaction is that all members of the *fragilis* group grow on media

Table 8. Groups of Bacteroides

	Fragilis	*Melaninogenicus-oralis*	Asaccharolytic
Bile salts	Growth	No growth	No growth
Glucose	+	+	–
Pigment	–	+*	–†

*A minority of strains do not produce pigment
†Except *B. asaccharolyticus* itself

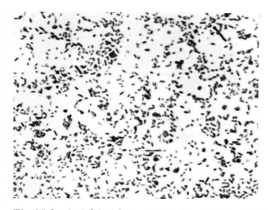 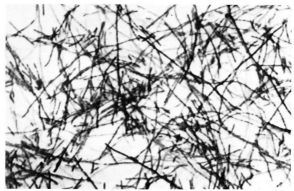

Fig. 75 On the left is a Gram-stained film of *B. fragilis* isolated from an infected appendicectomy wound in a diabetic patient. On the right a Gram-stained film of *B. necrophorus* showing its pleomorphic nature
×1000

containing sodium taurocholate and sodium deoxycholate which differentiates them from members of the *melaninogenicus-oralis* and asaccharolytic groups. Differentiation of the nine species of *B. fragilis* from each other requires detailed examination of fermentation reactions against a range of substrates and their varying capacities to produce indole, liquefy gelatin and hydrolyse aesculin.

Serological characteristics

Knowledge of the antigenic structure of *B. fragilis* species and other species within the genus is not yet sufficiently advanced to permit serological identification of species in service laboratories.

BACTEROIDES MELANINOGENICUS

The specific name was given to members of this group because colonies produce a black pigment derived from haemoglobin when grown on blood agar; this pigment is cell-associated and different from the

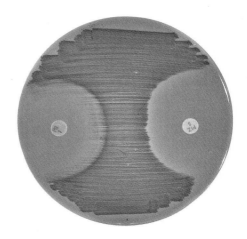

Fig. 76 Above is a culture of *B. fragilis,* below a culture of *B. melaninogenicus.* Although growth of both species is inhibited by metronidazole, the characteristic difference in response to penicillin is shown; note the black pigment typical of the appearance of *B. melaninogenicus* when growing on blood agar

pigmentation (ferrous sulphide) given by other bacteroides and several other bacterial species when these are grown on medium containing ferrous sulphate (Fig. 76). A minority of strains within the *melaninogenicus-oralis* group do not produce pigment on BA medium but in common with all members of the group that minority cannot grow on bile-containing media.

Further differentiation between members of the *melaninogenicus-oralis* group and those of the *fragilis* group is gained by noting that members of the former group are sensitive to neomycin discs containing 1000 μg of the antibiotic but *B. fragilis* strains uniformly show resistance to such a concentration of neomycin.

Differentiation of the seven species within the *melaninogenicus-oralis* group demands investigation of fermentative abilities and other biochemical tests analogous to these allowing speciation of the fragilis group.

ASACCHAROLYTIC BACTEROIDES

These form the least commonly isolated strains from clinical material and are distinct from other bacteroides by their inactivity against glucose or other carbohydrates; nor can they grow on bile-containing media. The three species presently recognised within the asaccharolytic group can be differentiated since *B. asaccharolyticus* produces pigmented colonies (black or dark brown) on BA and is resistant to kanamycin (1000 μg disc); *B. corrodens* are sensitive to a similar concentration of kanamycin and on primary culture the colonies become embedded in pits on the surface of the medium; the third species, *B. praeacutus,* does not produce pigment and although sensitive to kanamycin is tolerant to the presence of gentian violet, which inhibits the growth of the two other members of the asaccharolytic group.

As mentioned in the introduction to this chapter many laboratories report strains simply as 'Bacteroides species' whilst others attempt a primitive subdivision into *B. fragilis* when the isolate is resistant to penicillin (2 units/disc) and group the penicillin-sensitive strains as 'Bacteroides species'; such a subdivision is of no help to the clinician since metronidazole is the drug of choice in the clinical field. It is hoped that more and more diagnostic laboratories will, provided staff and finances are available, attempt a more detailed identification of *Bacteroides* isolates.

BACTEROIDES INFECTIONS

The essentially commensal nature of such species has been noted and indeed they may play an important part in protecting the alimentary tract by their production of acetic and other acids. One of the most common scenes where *Bacteroides* species are involved in endogenous infection is as a complication of colonic surgery, including appendicectomy; two features should be noted, firstly that until 10 or so years ago it was not uncommon for bacteriologists to issue a report on pus from an abdominal wound reading 'Many Gram-negative rods and pus cells seen microscopically. Growth — NIL'; this, or a similar type of report emphasises that the more fastidious anaerobic species were not being isolated either because the organisms had died during transmission to the laboratory or equally that our techniques of anaerobic culture were inadequate. The latter reason for failure has been eliminated by all self-respecting laboratories but problems in maintaining viability of such fastidious species during transmission to the laboratory still exist. Secondly, although *B. fragilis* as a species is the isolate most frequently made (75% or more) from wound infection following large bowel surgery, that species is small in proportion to all *Bacteroides* species in the large intestine; several hypotheses are being actively pursued to explain its supremacy as an endogenous infecting agent.

Bacteroides species, as opportunist pathogens, are encountered particularly in compromised patients where precipitating factors, such as the use of immunosuppressive drugs, steroid therapy or endocrine deficiency encourage infection.

In addition to infection following colonic surgery *Bacteroides* species are incriminated at many other sites, and also septicaemia due to such species is more common than was once thought. Here the introduction of better blood-culture media and techniques have also helped to throw such organisms into the limelight.

The hypothesis that *Bacteroides* and other bowel bacteria may produce carcinogens from dietary fats is being enthusiastically pursued in several centres.

Vibrios and Campylobacters

17

VIBRIOS

The genus *Vibrio* comprises three members pathogenic to man; other species are pathogenic for animals and insects and may lead a commensal existence in numerous hosts or occur as saprophytes, particularly in water.

VIBRIO CHOLERAE

Popularly known as the comma bacillus, this is the causative agent of classical Asiatic cholera.

Microscopy

When freshly isolated they appear as definitely curved rods with rounded ends, 3 μm $\times$ 0.5 μm. Often lose their curved appearance on artificial cultivation. Gram-negative. Motile with a single terminal flagellum. Non-capsulate and non-sporing.

Cultural appearances

Aerobic. Wide temperature range. Optimum 37°C. Grows on ordinary media but sensitive to acid pH; growth favoured by alkaline media and the optimum reaction is about pH 8.2. On agar the colonies are not distinctive, 2–3 mm in size after 18–24 h at 37°C, low convex with an entire edge, whitish and translucent. Older colonies develop a light ochre tint. Monsur's and Aronson's media are commonly employed for selective cultivation; by inoculating a tube of peptone water with a flake of mucus from the stool and incubating for only 6–8 h vibrios, if present, can be harvested from the surface in almost pure culture.

Biochemical abilities

V. cholerae ferments glucose, sucrose, mannitol and maltose without gas production but does not utilise lactose or dulcitol. It gives a positive *cholera-red reaction* when growing in peptone water due to the production of indole *and nitrites*. This can be tested by adding four drops of H_2SO_4 to a 72 h culture.

V. cholerae is non-haemolytic when 1 ml of a 2 day broth culture is added to 1 ml of a 5% suspension of sheep red cells (Greig test).

Serological characters

There are two recognised serotypes of *V. cholerae* — 'Inaba' and 'Ogawa' — each of which possesses a common H antigen but distinctive subsidiary O components.

VIBRIO EL TOR

This organism, which has been incriminated in epidemic situations, usually differs from the classical *V. cholerae* in being haemolytic in the Greig test; additionally, strains are resistant to Mukerjee's *V. cholerae* group IV phage and to polymyxin B (50 units/disc) whereas *V. cholerae* strains are sensitive.

V. el tor strains belong solely to the Inaba serotype.

VIBRIO PARAHAEMOLYTICUS

This is a halophilic (salt-loving) marine vibrio and its natural environment must be reflected by incorporating 2–4% NaCl in laboratory culture media. It is readily differentiated from *V. cholerae* and *V. el tor* since these species produce large yellow sucrose-fermenting colonies on thiosulphate bile sucrose (TCBS) agar whereas *V. parahaemolyticus* colonies are blue-green in appearance on TCBS medium.

CHOLERA

This disease, which was once widespread throughout the world, has now been contained in its original home in Asia; its absence from most other continents depends on the continued provision of filtered and chlorinated water which, after such treatment, is stored and delivered so that it cannot be contaminated.

Like typhoid fever, cholera is essentially a water-borne infection and the sources from which water supplies can be contaminated are cases of the disease; transient carriage can occur after recovery but carriage beyond a few weeks is rare.

Epidemic spread is frequently associated with the holding of religious and other festivals, when the meagre sanitary arrangements break down, and also in these circumstances case-to-case infection probably occurs via contaminated fomites and food.

Although El Tor vibrios can be distinguished from *V. cholerae* strains in the laboratory, the clinical syndrome and the sources and modes of spread of cholera caused by the former organisms are identical with those of classical cholera.

Although active immunisation is practised there is no proof that this is protective for perhaps more than 3–6 months. Emergency preventive measures involve temporary chlorination of water supplies and the boiling of all water to be used for domestic purposes. On a long-term basis sanitary engineering has more to offer than medicine in the conquest of cholera.

In contrast to the above picture *V. parahaemolyticus* is associated with food poisoning; infection occurs from eating shellfish or raw fish which inhabit warm coastal waters, e.g. in Japan. The incubation period usually is within 8–18 h and spectacular epidemics occur from time to time when air travellers become infected en route to Britain. Human to human transmission is rare.

CAMPYLOBACTERS

Species now allocated to the genus *Campylobacter* were formerly classified as vibrios and one, *C. foetus,* has long been recognised as a cause of abortion in cattle and sheep but only rarely as an opportunist pathogen in compromised human hosts. In 1977, *C. jejuni* was recognised as a popular pathogen of the human intestinal tract and is now accepted, globally, as one of the most significant bacterial intestinal pathogens.

CAMPYLOBACTER JEJUNI

Microscopy

Identical to *V. cholerae* but with amphitrichous flagella.

Cultural appearances

The somewhat unusual conditions required for isolation of this species probably accounts for its recent recognition as a human pathogen. *Microaerophilic* conditions are essential and optimally an atmosphere of 5–10% O_2 and 3–10% CO_2 must be provided, the balance of the atmosphere comprising an inert gas such as hydrogen; likewise *C. jejuni* grows happily at 43°C. *C. foetus* strains thrive at 25°C whilst *C. jejuni* strains do not.

In addition selective media are used which allow *Campylobacter* species to grow but discourage the growth of most other organisms in faecal specimens; selectivity is usually acquired by incorporating antibiotics in the medium, e.g. vancomycin, trimethoprim and polymyxin B.

Colonies are low convex, glossy and effuse and have been likened to droplets of water.

Biochemical abilities

Inability to grow at 25°C but rapid growth at 43°C differentiates *C. jejuni* from *C. foetus*. Sensitivity to nalidixic acid serves further to differentiate *C. jejuni*.

Other tests, e.g. ability to grow in the presence of 1.5% NaCl and sensitivity to metronidazole may prove useful in more detailed subdivision of the genus.

Serological characters

A serotyping system has been devised but since strains from a given epidemic belong usually to the same serotype further exploration is required to assess its validity. Antibodies to the infecting strain can be demonstrated in the patient's serum by complement fixation, agglutination or bactericidal testing methods.

CAMPYLOBACTER INFECTIONS

Numerous natural hosts, other than man are known to excrete *C. jejuni*, including cattle, birds, sheep and pigs; dogs and cats are also commonly infected.

Most reports of *C. jejuni* epidemic infections in humans in the UK have been associated with drinking raw, unpasteurised milk.

Case to case transmission in the human host is unusual in adults but does occur when the patient is a young child who becomes faecally incontinent as a result of infection.

The carrier rate of *C. jejuni* in the human host appears to be of a low order and infection seems almost certainly zoonotic; on recovery from infection carriage is brief and a chronic carrier state has yet to be described.

The main presenting features after a 2–5 day incubation period are diarrhoea and abdominal pain and during the incubation period a brief flu-like illness is often experienced.

Although the illness is often referred to as campylobacter enteritis it should be appreciated that the large bowel is also usually infected; cases vary in clinical severity but as with all other bacterial gut infections (*except* typhoid fever) antibiotic treatment is rarely indicated in campylobacter disease.

In rare instances of severe symptomatology, erythromycin is presently the antibiotic most likely to assist recovery.

At present *Campylobacter* infections in the UK and many other temperate zone countries, are rivalling salmonellas and shigellas as the most common bacterial bowel infection; it will be interesting to see if *C. jejuni* retains this position or whether, after the early excitement of this new disease, the longer recognised bacterial bowel pathogens will reassert their authority in the human population.

A point of interest is that Koch's postulates have been fulfilled, since volunteers who consumed *C. jejuni* cultures showed all the characteristics of the infection and the same serotypes as were ingested were recovered in large numbers from their faeces; they also developed significant titres of specific antibody.

Haemophili

18

Members of this genus are strictly parasitic in man or other animals and are feebly viable outside their respective hosts and, for growth, require to be supplied with certain constituents of blood (haemophilic). *Haemophilus influenzae* is the commonest member of the genus pathogenic to man, with *H. ducreyi* less commonly found.

HAEMOPHILUS INFLUENZAE

This organism is parasitic only on mankind, and is frequently found as a commensal in the oropharynx but has potential as a lethal organism not only in the respiratory tract but in acute meningitis in pre-school children.

Microscopy

Characteristically small cocco-bacilli 1.5 μm × 0.5 μm. Gram-negative, non-motile, capsulate in young cultures, non-sporing. Often presents as very long filaments, particularly in cerebrospinal fluid from cases of haemophilus meningitis and also in old laboratory cultures (Fig. 77).

Cultural appearances

Aerobic and demands blood enriched media

Fig. 77 Gram-stained film of a young culture of *H. influenzae;* coccal and cocco-bacillary Gram-negative forms with no evidence of pleomorphism which characterises older cultures
× 1200

for growth; blood agar or, more so, chocolate blood agar provides the X (haematin) and V (diphosphopyridine nucleotide) factors required for growth; heating of the blood to produce chocolate blood agar not only releases V factor from the red cells but also inactivates V-factor destroying enzymes present in the blood cells. On these media, colonies are small (1–2 mm) and transparent; *H. influenzae* grows in symbiosis with staphylococci which produce V factor. Hence in the vicinity of staphylococci, colonies of *H. influenzae* are larger in size, a feature referred to as 'satellitism' (Fig. 78).

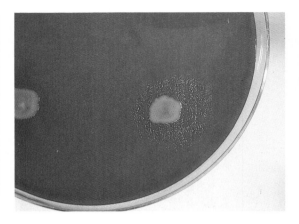 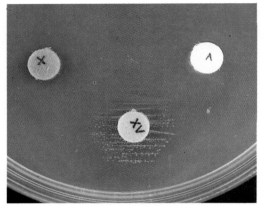

Fig. 78 On the left, satellitism is evident by the enhanced growth of *H. influenzae* around the solitary colony of *Staph. pyogenes;* the dependence of *H. influenzae* on both X and V factors is shown by the absence of growth round the disc containing the separate factors compared with growth in the area around the combined disc; the medium is nutrient agar, which does not support the growth of *H. influenzae* without these essential growth factors

Dependence or otherwise on X and/or V factors within members of the genus is readily demonstrated by sowing a plate of nutrient agar with the organism to be investigated and placing three discs containing X factor, V factor and a combination of both factors widely separated one from the other on the surface of the seeded plate before incubation. *H. influenzae* will only grow around the combined (X and Y) disc; *H. parainfluenzae* will grow round the combined disc and also round the V-factor disc.

Biochemical activities

With the important proviso that suitable fluid media, e.g. Levinthal broth, are used, *H. influenzae* strains grow rapidly and a biotyping scheme allows recognition of six types (Table 9).

Table 9. Types of Haemophilus influenzae

Biotype	Indole production	Urease production	Ornithine decarboxylase
I	+	+	+
II	+	+	−
III	−	+	−
IV	−	+	+
V	+	−	+
VI	−	−	+

Biotype I strains encompass the vast majority of isolates from severe clinical infections; this recently introduced method of typing has still to be fully evaluated *vis à vis* serotyping in regard to epidemiological facts but appears promising.

Serological characters

Smooth strains possess capsular polysaccharide antigens designated more than 50 years ago by Pittman as types a–f and the six types can be recognised by a capsule swelling reaction analogous to that employed in typing pneumococci and klebsiellae.

Animal inoculation

Not pathogenic for laboratory animals

HAEMOPHILUS PARAINFLUENZAE

Such strains are very similar to *H. influenzae* in virtually all respects except that *H. parainfluenzae* requires only the V factor for growth and some isolates produce β-haemolysis on blood agar.

HAEMOPHILUS DUCREYI

Strains resemble *H. influenzae* but require only the X factor for *in vitro* growth. *H. ducreyi* causes a sexually-transmitted infection known as chancroid or soft sore; the organism can be seen in films made from the primary lesion and from material aspirated from the swollen associated regional lymph glands. Confirmation of identity finally rests with agglutination tests with a specific antiserum.

HAEMOPHILUS HAEMOLYTICUS

Virtually identical in its biological characteristics with *H. influenzae*, including its need for both X and V factors for growth; such strains however produce β-haemolysis when grown on blood agar plates and since *H. haemolyticus* is a normal inhabitant of the human upper respiratory tract colonies might be readily mistaken for those of β-haemolytic streptococci unless the observer undertakes the elementary precaution of performing a Gram stain on such colonies.

HAEMOPHILUS INFLUENZAE INFECTIONS

H. influenzae can be isolated from the nasopharynx of 50–80% of people and appears to fulfil a commensal role in most instances; the majority of such strains are rough, non-capsulate forms and are found in association with localised non-bacteraemic infections and usually in adults, e.g. exacerbations of chronic bronchitis, sporadic cases of sinusitis, conjunctivitis and otitis media.

On the other hand strains causing meningitis or other acute infections have a low carrier rate (2–4%) and are almost invariably of Pittman's capsular type b and are of biotype 1; haemophilus meningitis is most common in children of pre-school age and *H. influenzae* is rarely encountered as a cause of acute bacterial meningitis after the fifth birthday. The unusual morphology of *H. influenzae* in the CSF of cases of meningitis has already been noted.

Epiglottitis is another acute manifestation of infection with *H. influenzae*, and here too type b strains predominate; this acute inflammation of the epiglottis and other supraglottic structures occurs usually in young children (usually 2–4 yr) and although, as in haemophilus meningitis, there is an accompanying septicaemia, the usually sudden onset of marked local oedema can of itself by lethal by asphyxiation.

Less common or less alarming infections caused by *H. influenzae* are pneumonia, suppurative arthritis and osteitis, otitis media and facial cellulitis. In all of these capsulate type b strains predominate.

Epidemics of acute haemophilus conjunctivitis occur in some tropical countries and until recently the causal organism was given specific status, i.e. *H. aegyptius;* these are now allocated to biotype III of *H. influenzae*.

Finally it is worth recalling that the specific epithet *influenzae* was originally associated with the prospect that such strains were the cause of influenza, and although the viral origin of this infection is beyond all doubt, bacteria, including *H. influenzae* strains, have a role to play as secondary pathogens in viral influenza, on occasion with devastating results even in the previously healthy young individual.

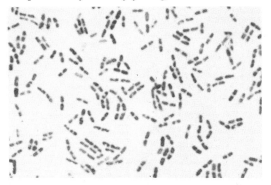

Fig. 79 Gram-stained film of *Moraxella lacunata*, a short, thick, Gram-negative diplobacillus. This organism was previously classified with the haemophili but does not require accessory growth factors. Its specific name is associated with its ability to liquefy blood agar medium so that the colonies float on fluid in lacunae or craters in the medium

× 1200

Bordetellae

19

The generic title commemorates the isolation by Bordet of the whooping-cough bacillus, *Bordetella pertussis;* this organism, along with two closely related species, *Bord. parapertussis* and *Bord. bronchiseptica,* was originally classified with *Haemophilus* but since bordetellae require neither X nor V factors for growth they are now given separate status.

BORDETELLA PERTUSSIS

Microscopy

Bears a close resemblance to *H. influenzae* but is less pleomorphic.

Cultural appearances

Complex enriched media, such as Bordet-Gengou's potato blood glycerol agar, are required for primary isolation; even then, 2–3 days incubation at 37°C are required before colonies can be recognised. These are small, dome-shaped and highly refractile.

Biochemical activities

No reliable or distinctive fermentative properties. Does not produce urease, whereas other *Bordetella* species do so.

Serological characters

All strains of *Bord. pertussis* possess a common surface antigen, and in addition one or more additional surface antigens; thus serotypes isolated from cases of infection may have differing antigenic components and we refer to these as *Bord. pertussis* type 1, 3 or 1, 2, 3 or type 1, 2. It is important that the strain used for vaccine production should possess the antigens relevant to the strains causing whooping cough in a given community at a given time.

BORDETELLA PARAPERTUSSIS

This species is similar in many respects to *Bord. pertussis* but when growing on Bordet-Gengou medium, colonies produce darkening of the blood underlying them. Urease producer. Strains are serologically homogeneous and although there are antigenic components in common with *Bord. pertussis,* strains of *Bord. parapertussis* can be agglutinated with an absorbed antiserum.

Bord. parapertussis strains alone are isolated in a small proportion of cases of whooping cough. Essentially a commensal of the human pharynx.

BORDETELLA BRONCHISEPTICA

Infection with this species is rare in man but animals, particularly rodents, are the natural host. *Bord. bronchiseptica* is the only motile member of the genus and possesses peritrichous flagella. Grows readily on nutrient agar. Produces urease. Whilst serologically related to other bordetella strains it can be specifically identified using absorbed antiserum.

WHOOPING COUGH

Whooping cough, or pertussis, is restricted to humans and is essentially a disease of childhood; although the fatality rate is low, approximately 50% of deaths occur in those infected in the first six months of life.

It is probable that infection is spread by direct droplet spray from an infected individual but the relative importance of the part played by other methods of spread such as indirect contact via formites and dust-borne spread is not yet known.

The disease is most highly communicable in the early catarrhal stage before the typical paroxysmal cough or whoop appears; thereafter communicability decreases fairly rapidly so that within 2–3 weeks of the onset of whooping the case is no longer a source of infection, although coughing may continue for several more weeks.

Whooping cough is one of the few infections which occur more frequently and are more severe in females than in males; in communities unprotected by active immunisation 80% or more of the children suffer infection and the disease recurs in epidemic fashion every three to four years.

Active immunisation with a potent pertussis vaccine should be undertaken in the first six months of life beginning at the second month; in this way we can influence the incidence of the disease at a time when the fatality rate is highest.

In comparison with a 90% attack rate among unimmunised siblings, the attack rate for protected children with home exposure is, with a potent vaccine, less than 10%.

Whooping cough does occur in immunised children, but the disease is very much milder and of much shorter duration than in unprotected children.

The introduction in 1955 of a competent vaccine for mass immunisation was rapidly followed by a steady reduction in the incidence and mortality of pertussis, then in 1974 political and public debate arose concerning the risks of the vaccine, which is usually incorporated with toxoid preparations to give simultaneous protection against diphtheria and tetanus, the Triple Vaccine.

The main concern was the allegedly unacceptable incidence of encephalopathy and occasional permanent brain damage in immunised infants; there is still debate on the safety of the vaccine but the rapid drop in acceptance rates has seen a rebound in incidence of whooping cough to levels commonly encountered before 1955.

Detailed ongoing surveys in the Greater London area and in Glasgow suggest that if the few children who might be at greater risk from post-immunisation encephalopathy are not given the triple vaccine then it remains a safe agent with excellent powers.

Children who should NOT be given pertussis vaccine include those with a history of convulsions, those in whom a first-degree relative has a history of epilepsy and those who at the time for immunisation are suffering other infection; likewise children who have suffered any ill effect from a prior dose of vaccine should not receive further doses.

It is not sufficiently appreciated that a double toxoid (diphtheria and tetanus) vaccine is available for those children whose parents wish to avoid the possible risk of the pertussis component of the triple vaccine; tragically the rapid drop in acceptance rates of triple vaccine have been mirrored by reduced acceptance of immunisation against poliomyelitis and we are, as a nation, in danger of creating a young population increasingly at risk from essentially preventable infections.

Yersiniae, Pasteurellae and Francisellae

20

YERSINIA

Members of this genus were until recently classified as pasteurellae but the taxonomists tell us that they properly belong to the family Enterobacteriaceae; the most notorious member is *Yersinia pestis,* the plague bacillus, which still haunts certain areas of the world although it has long been eradicated from most developed countries including Britain. The other members of the genus affecting man are *Yersinia enterocolitica* and *Yersinia pseudotuberculosis* and, like the plague bacillus, their natural hosts include rodents.

YERSINIA PESTIS

Y. pestis has in the past decimated the world population on several occasions when it has been transmitted to man by the bite of the rat flea *Xenopsylla cheopis.*

Microscopy

Short ovoid bacilli, 1.5 μm × 0.7 μm, Gram-negative, non-motile, capsulate in tissues and on primary *in vitro* isolation. Non-sporing. When stained, the central part of each bacillus is often uncoloured to give the appearance of bipolar staining. Pleomorphism is a constant feature on prolonged cultivation or after several subcultures *in vitro* (Fig. 80).

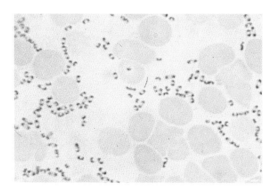

Fig. 80 *Yersinia pestis* in a blood film stained by Leishman's method; the bipolar staining is characteristic × 2000

Cultural appearances

Aerobic and facultatively anaerobic. Optimum growth temperature is unusual, i.e. 30°C. Colonies are minute after 24 h, pinpoint and semi transparent; continued incubation, e.g. 48–72 h, gives larger colonies with an irregular perimeter (Fig. 81).

Biochemical activities

These are shown in Table 10 and contrasted

Table 10. Cultural appearance of yersiniae

	Y. pestis	*Y. enterocolitica*	*Y. pseudotuberculosis*
Growth at 5°C	−	+	−
Urease production	−	+	+
Sucrose	−	+	−
Aesculin	+	−	+
Motility*	−	+	+

*Tested at 22°C

Fig. 81 48 h culture of *Y. pestis* incubated at 30°C. Colonies are small, greyish and semitransparent; this particular strain, unlike the majority, could not be grown on MacConkey's medium in spite of being biochemically and serologically proven to be *Y. pestis*

with those of *Y. enterocolitica* and *Y. pseudotuberculosis.*

Serological characters

Strains are serologically homogeneous and although two toxins, one associated with the cytoplasmic membrane and the other within the cytoplasm, have high lethality for experimental animals, these same toxins are also present in avirulent strains; thus virulence must depend on factors in addition to these two toxins.

Animal inoculation

In addition to its natural host, the rat, many other rodents are susceptible. Subcutaneous injection of a culture into a white rat causes death within a few days and autopsy reveals local oedema and necrosis at the inoculation site; in addition, the related lymph glands are markedly enlarged as in the spleen. *Y. pestis* is recoverable from these sites and also from the heart blood in the terminal stages of the illness.

YERSINIA ENTEROCOLITICA

In comparison with the plague bacillus, *Y. enterocolitica* is a new organism, since it was first isolated in 1949, some 55 years after *Y. pestis* was discovered independently by Kitasato and by Yersin during the Hong Kong plague epidemic in 1894!

Microscopy

Features are similar to those of *Y. pestis* BUT displays motility by virtue of peritrichous flagella when incubated at 22°C.

Cultural appearances

Growth requirements are similar to those of *Y. pestis* but it should be noted that *Y. enterocolitica* strains grow (although slowly) at 5°C. This virtue was used for isolation of strains from faecal specimens until recently; nowadays selective isolation is more readily obtained by treating faecal specimens with 0.5% KOH before plating out on MacConkey's medium.

Biochemical activities

Relevant tests to differentiate *Y. enterocolitica*

109

from other species within the genus are given in the previous Table.

Serological characters

Numerous types based on different somatic antigens are recognised but such typing is undertaken only by a few reference laboratories.

YERSINIA PSEUDOTUBERCULOSIS

Microscopically identical to *Y. enterocolitica* especially in being motile at 22°C. Other contrasting features are shown in the Table.

PLAGUE

Plague is essentially an epizootic infection in wild rats and certain other rodents; man is affected only in circumstances where his environment allows the vector rat flea easy access to him, e.g. overcrowding in insanitary living quarters.

Bubonic plague is the type most commonly seen in man and results from the bite of an infective rat flea; the rat flea becomes infected by taking a blood meal from a rat and the bacilli then multiply in the stomach and proventriculus. Then, if and when the flea attempts to take a blood meal from man, plague bacilli are regurgitated from the proventriculus into the tissues.

A few days after being inoculated by the flea the patient shows progressive swelling of the regional lymph glands and surrounding tissues – this lesion is known as the primary bubo; secondary buboes may occur in other lymph glands. The buboes are packed with plague bacilli.

In bubonic plague infection rarely spreads from person to person but if a septicaemic phase develops then a rapidly fatal broncho-pneumonia occurs and this *pneumonic plague* can spread to other healthy people by airborne

routes. Since *Y. pestis* can survive for several weeks on sputum-contaminated surfaces, pneumonic plague may be dust-borne as well as being spread by droplet nuclei.

Prevention of plague depends on rodent control and destruction of fleas; during epidemics individuals at special risk must wear protective clothing and masks. Medical and nursing personnel should be given prophylactic doses of broad-spectrum antibiotics.

YERSINIOSIS

The term yersiniosis would suggest a specific syndrome similar for example to shigellosis; however human infections caused by *Y. enterocolitica* and *Y. pseudotuberculosis* vary from severe septicaemic-type illness similar to typhoid fever through to mild enterocolitic infections with non-bloody diarrhoea to mesenteric lymphadenitis which presents as appendicitis.

Recently such organisms are being suspected of association with Crohn's disease.

Wild and domesticated animals act as reservoirs internationally and it is probable that human infections from such zoonotic sources are largely overlooked.

PASTEURELLAE AND FRANCISELLAE

The genus *Pasteurella* now comprises only one human pathogen *Pasteurella multocida (P. septica)*, although many other species within the genus are pathogenic to a wide variety of animals and birds and thus have an indirect influence on man's economy.

P. multocida has many similarities to the Yersiniae but does not grow on MacConkey's medium and is non-motile regardless of the temperature of incubation; it does not produce urease but does produce indole.

Human infections in Britain are usually mild and almost invariably result from a bite or scratch from dogs and cats.

Francisella tularensis shares several features of *Yersinia and Pasteurella* species, particularly in being a natural pathogen for various rodents and affecting man only incidentally. In the case of *F. tularensis,* infection from rabbits and hares is an occupational hazard and the disease is spread by lice, mites and ticks from the natural rodent host to trappers or others handling infected animals.

Tularaemia seems to be restricted to certain countries and has an especial association with the USA where the human disease was first described in 1911.

F. tularensis has several features in common with *Y. pestis,* including bipolar staining; requires complex media for growth and even on these, colonies are slow to develop and identification depends primarily on the demonstration of agglutination of the isolate with specific antiserum.

Brucellae

21

All members of this genus are strictly parasitic on man or animals and are characteristically intracellular. Brucellosis in the human subject is classically associated with drinking unpasteurised goat's milk (*Brucella melitensis*) (Figs. 82 and 83) or cow's milk (*Br. abortus*). In either case and also with *Br. suis* (affecting pigs) there is an occupational hazard to farm workers, abattoir personnel and veterinary surgeons. Laboratory workers may become infected from handling cultures.

BRUCELLA ABORTUS

Microscopy

Varies in shape from coccal forms (0.5 μm) to small bacilli (1 μm $\times$ 0.5 μm). Gram-negative. Non-motile. Capsulate. Non-sporing.

Cultural appearances

Aerobe. Grows on ordinary media but a more

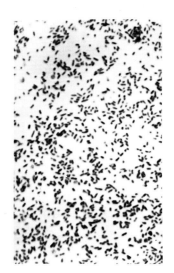

Fig. 82 Gram-stained film of *Br. melitensis* isolated in Britain from a person recently returned from the Middle East who had, for a few weeks, been addicted to drinking raw goat's milk. Gram-negative coccal and cocco-bacillary forms are seen
  $\times$ 1000

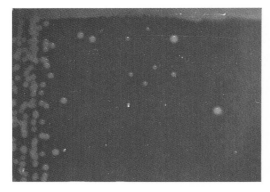

Fig. 83 Colonies of *Br. melitensis* isolated on blood agar after 48 h incubation at 37°C. Culture was from the same case as in Fig. 82, and has the features noted in the above paragraph
  $\times$ 2.25

112

Fig. 84 Dye inhibition tests. Basic fuchsin at a concentration of 1: 25 000 (left) inhibits only the growth of *Br. suis* (the lowest of the three growth bands) whereas thionine (1: 30 000, right) inhibits only the growth of *Br. abortus* (central growth band). Neither dyestuff, at the concentration stated, inhibits strains of *Br. melitensis*, which is the upper growth band

rapid and luxuriant growth is obtained by cultivation on liver-infusion agar. Colonies are low convex, with an entire edge, transparent and approximately 1 mm in diameter; on continued incubation they increase in size and become brownish.

Biochemical abilities

In ordinary sugar media no fermentation is observable, but if a peptone-free, buffered medium containing the relevant substrate is heavily inoculated, very consistent patterns are obtained.

Serological characters

Br. abortus is very closely related to the other members of the genus since all possess two similar antigens; however, one of these latter is dominant in *Br. abortus,* and thus it is possible to prepare an absorbed agglutinating antiserum for this species.

Animal inoculation

Guinea-pigs inoculated intramuscularly or subcutaneously with *Br. melitensis* suffer a chronic illness which is rarely fatal and differs

Table 11. Differential characteristics of brucellae

	Br. abortus	*Br. melitensis*	*Br. suis**
Glucose	$\perp$	$\perp$	$\perp$
Inositol	$\perp$	–	–
Maltose	–	–	$\perp$
Inhibited by:			
a. Basic Fuchsin 1:25 000	No	No	Yes
b. Thionine 1:30 000	Yes	No	No
CO_2 required for growth	Yes	No.	No.
H_2S produced	Yes	No	Yes

*American strains. Danish strains are similar except for inability to produce H_2S.

from that caused by inoculation with other brucellae.

The features which differentiate *Br. abortus* from the other members of the genus, *Br. melitensis* and *Br. suis* are given in Table 11.

BRUCELLOSIS

Brucellosis is primarily an infection of certain animals, and in Britain *Br. abortus* is the species most commonly incriminated in bovine and human infections.

Whilst the general population is at risk of infection if unpasteurised milk from an infected cow is drunk, there are certain occupations with a high risk of acquiring infection by coming in contact with sick animals or their discharges. Similarly the handling of infected carcases allows the organisms to pass through skin abrasions or to be inhaled; abortion is a common phenomenon in infected cattle but is rare in infected pregnant women and it is thought that this is explained by the fact that whereas bovine and other animal placentae contain erythritol which stimulates the growth of brucellae, erythritol is absent from the human placenta.

The severity and duration of brucellosis in man are subject to wide variation and the disease should be considered as a possible cause when cases of pyrexia of uncertain origin are being investigated. As with almost all other zoonoses, infection rarely spreads from man to man.

Prevention of brucellosis in those at occupational risk demands the education of farmers, abattoir workers and others handling carcases in methods of minimising the risk and particularly in the manipulation and disposal of discharges and foetuses from aborted animals.

Whilst the prevention of brucellosis in the general population is dependent on the eradication of the disease in the animal population, pasteurisation of all milk offers an effective safeguard against acquiring infection by this vehicle.

In several countries *control* of infection in the animal has been by vaccination of calves and the simultaneous segregation of infected adult animals; *eradication* began in Britain a few years ago and large areas of the country are now stocked with *Brucella*-free herds. Eradication policies required the detection and sacrifice of infected animals, a policy analogous to that used decades ago to build up cattle herds free from bovine tuberculosis; such a policy requires not only cooperation from the farmers but is costly; however the final saving both to the farmer and from the elimination of human brucellosis is immeasurable.

Such eradication schemes ultimately rely on honesty, since the unscrupulous cattle dealer who smuggles cattle into an area free from brucellosis can easily undo years of cooperation if even one illicitly imported beast is suffering from the disease.

Recent activities of this kind have caused a partial breakdown in the maintenance of tubercle-free herds in Ulster.

Borreliae

22

Members of this genus are pathogenic or potentially pathogenic for man or animals. As a genus they differ from other spirochaetes in being larger and therefore visible with the compound light microscope after staining with the normal dyes.

BORRELIA VINCENTII

This is the only member of the genus found in

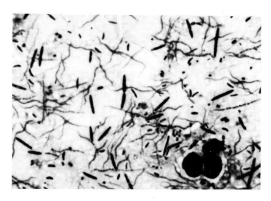

Fig. 85 Film from a case of gingivo-stomatitis after staining for 10 min with dilute (1 : 10) carbol fuchsin. Large numbers of *Borr. vincentii* and fusiform bacilli *(Fusiformis fusiformis)* are seen. Our inability to grow these species easily demands that, at present, we rely on such microscopic appearances for diagnosis.

Vincent's infection responds rapidly to penicillin therapy.

× 1000

Britain; it leads a commensal existence in the healthy human mouth but when the mucous membrane is devitalised, e.g. by trauma or nutritional deficiency, *Borr. vincentii* may assume a pathogenic role in association with fusiform (cigar-shaped) bacilli. Vincent's infection is frequently seen as an ulcerative gingivo-stomatitis and occasionally as ulcerative tonsilitis. These symbiotic organisms are sometimes found in lung abscesses and cases of bronchiectasis.

The diagnosis of Vincent's infection depends on the microscopic findings of *large numbers* of fusiform bacilli and *Borr. vincentii;* the latter is 5–20 μm × 0.3 μm and possesses three to eight irregular spirals; Gram-negative and non-sporing. It is strictly anaerobic and difficult to cultivate (Fig. 85).

BORRELIA RECURRENTIS

The causal organism of European relapsing fever.

Microscopy

10–30 μm × 0.3 μm with five to seven fairly regular coils. Gram-negative and non-sporing. Highly motile (Fig. 86).

115

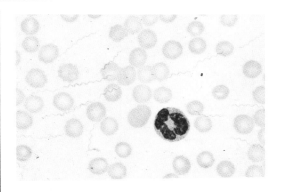

Fig. 86 Blood film from a case of relapsing fever. Several *Borr. recurrentis* are seen and are morphologically identical with other species involved in relapsing fever in various countries
  Leishman's stain; × 800

Cultural appearances

The organism is very difficult to grow even in heavily enriched fluid media. Not grown on solid media.

Biochemical abilities

No knowledge available.

Serological characters

No detailed knowledge but antiserum developed against *Borr. recurrentis* does not agglutinate *Borr. duttonii,* which is responsible for West African relapsing fever.

Animal inoculation

White mice, but not guinea-pigs, can be infected by subcutaneous inoculation of blood from a human case of relapsing fever; the organism can be demonstrated in stained films of tail blood 1–3 days later.

BORRELIA DUTTONII

Identical with *Borr. recurrentis* except for its serological characters; it is antigenically distinct from all other relapsing fever spirochaetes.

Many other Borreliae have been associated with relapsing fever in different parts of the world and specific names have been given to them often on flimsy evidence.

RELAPSING FEVERS

European relapsing fever is louse-borne from man to man by *Pediculus humanus* var. *corporis;* after the louse has taken a blood meal from an infected individual the spirochaetes can be demonstrated in the louse's stomach for 12–24 h and then they disappear. Some 5–7 days later they reappear throughout the louse's body and man can then be infected by crushing lice on his skin whilst scratching and thus introducing the spirochaetes through abrasions; alternatively the insect may bite the patient and the bite-wound can then be contaminated with the louse's infective excreta.

Prevention depends primarily on delousing of individuals, their clothing and the household environment.

West African relapsing fever is tick-borne, and the natural reservoirs of infection in this type of relapsing fever are wild rodents and also the ticks themselves, since the spirochaetes are transmitted transovarially to consecutive generations of ticks.

Obviously the control of tick-borne relapsing fever is more difficult than that of louse-borne infection since mammals other than man are involved in the cycle of infection, and also, ticks once infected may remain infective for several years.

Treponemata

23

There are numerous species within this genus but only a few are pathogenic to man and of these latter only *Treponema pallidum* is found in Britain; many of the commensal species occur in situations, e.g. on the genitalia, where their differentiation from *Tr. pallidum* is of importance in the diagnosis of syphilis.

TREPONEMA PALLIDUM

The causative organism of syphilis

Microscopy

Delicately spiralled filaments 6–14 μm × 0.1 μm with 6 to 12 small regular coils. Ends are pointed and never recurved. Feebly refractile and requires dark-ground illumination for its demonstration in the unstained state, e.g. in material from the primary chancre. Not amenable to ordinary staining methods but silver impregnation stains may be used. Immunofluorescent techniques allow visualisation of the organism in tissues.

Cultural appearances

Tr. pallidum has not been cultivated in artificial media or in cell culture but can be maintained by intratesticular inoculation in the rabbit, an expensive and time-consuming process.

Our inability to grow the organism *in vitro* explains our present ignorance of most of its biological characteristics.

Serological characters

For diagnostic purposes it is sufficient to realise that infection of humans stimulates the formation of two categories of antibody; one of these categories is stimulated by cardiolipin which is present in the lipid fraction of treponemes and tests for antibodies to cardiolipin are non-specific and detect antitreponemal antibodies. The earliest of such tests was introduced by Wassermann in 1906. Other non-specific anti-cardiolipin tests, e.g. the Venereal Diseases Research Laboratory (VDRL) test are still useful in prognosis when the patient is being treated.

In contrast, tests to detect *specific antibodies* for pathogenic treponemes use as antigen either intact *Tr. pallidum* (maintained by intratesticular inoculation of rabbits) — the Treponema pallidum immobilisation test (TPI) — or extracts of *Tr. pallidum* as in the fluorescent treponemal antibody absorbed test (FTA-ABS); or the Treponema pallidum haemagglutination test (TPHA), where the test agent is sheep r.b.c.s. coated with an extract of *Tr. pallidum,* and the coated red cells are agglutinated if specific antibody is present in the patient's serum.

TREPONEMA PERTENUE

The cause of yaws, a non-venereal but communicable disease in tropical countries. Identical with *Tr. pallidum* but originally regarded as more slender, hence the specific name.

In Arabia, another non-venereal disease, bejel, and in tropical America and certain areas of the Pacific the disease pinta, are both apparently caused by organisms indistinguishable from *Tr. pallidum;* the specific name *Tr. carateum* is used to describe the organism associated with pinta.

In all these non-venereal treponematoses the serological responses do not allow differentiation from syphilis but the clinical presentations are distinctive.

COMMENSAL TREPONEMATA

These are found on the genitalia or in the mouth; *Tr. gracile* and *Tr. genitalis* in the former situation and *Tr. microdentium* in the latter. Microscopic differentiation from *Tr. pallidum* is difficult and must be undertaken with great care; these commensals can be grown in fluid media under strictly anaerobic conditions. These surface commensals can be avoided if care is taken to prepare the area of the suspect syphilitic sore by thorough cleansing before collecting serous exudate.

SYPHILIS

With few exceptions syphilis is contracted by sexual intercourse; *Tr. pallidum* is so feebly viable outside host tissues that infection acquired other than by sexual intercourse *always* involves direct contact, e.g. manual infection in doctors and nurses who have examined carelessly a syphilitic lesion. One reflection of our modern society is that almost half of all new cases of primary syphilis each year now present as anorectal lesions in male homosexuals, so that even the most mundane examination per rectum can result in innocent infection of medical and nursing personnel.

The primary lesion or chancre appears 3–6 weeks after exposure to infection and is at first papular, but necrosis occurs with the formation of an indurated ulcer and *Tr. pallidum* is present in large numbers in the primary chancre which heals with the formation of a scar.

Secondary stage lesions appear 6–12 weeks after the appearance of the chancre and during this stage the spirochaetes spread throughout the patient's tissues via the bloodstream; generalised skin rashes, condylomata of the anus and vulva and mucous patches in the mouth are frequently seen in the secondary stage. All of these lesions contain large numbers of the causal spirochaete but their inability to survive on clothing, etc fortunately protects innocent contact with the patient.

Tertiary syphilis is characterised by the appearance of gummata in various organs some years after infection unless treatment has eliminated the infection; lesions in the central nervous system give rise to characteristic clinical syndromes, e.g. tabes dorsalis.

The control of sexually-acquired syphilis is similar to that of all venereal diseases and includes health and sex education, the provision of clinics for rapid diagnosis and treatment, the intensive follow-up of possible sources of infection and the screening and, if need be, treatment of contacts.

The ease with which gonorrhoea can be treated outside venereology departments carries the risk of masking concomitant primary syphilis, thus all cases of gonorrhoea should be kept under serological surveillance to ensure that syphilis was not acquired simultaneously.

Congenital syphilis has diminished in incidence in recent years and much of the reduction can be attributed to careful antenatal care of the pregnant woman which must include serological tests to exclude the possibility that she is suffering from syphilis; alternatively intensive antibiotic therapy of the syphilitic expectant mother may reduce the chances of her child suffering infection.

Leptospirae

24

More than 160 pathogenic members of the genus have been recognised to date and these vary in their geographic distribution and in their host range; rats, dogs, field mice, hedgehogs, cattle and pigs as well as other animals, are natural hosts for different leptospires and man may be incidentally parasitised.

Taxonomic changes have been at play in recent years and it is now recommended that only one species be recognised, i.e. *Leptospira interrogans* comprising two complexes, namely 'interrogans' (parasitic species) and 'biflexa' (saprophytic species); parasitic species are subtitled on a serological basis, thus *L. interrogans* serovar. *icterohaemorrhagiae* is now the accepted title of *L. icterohaemorrhagiae,* the cause of Weil's disease in man.

Such taxonomic exactness may confuse the clinician, and many bacteriologists continue to report using the older nomenclature and use more detailed serotyping solely for epidemiological purposes.

PATHOGENIC LEPTOSPIRES

These are identical in their morphology, cultural requirements and appearances as well as their biochemical behaviour and can be differentiated only be antigenic analysis.

Microscopy

Finely coiled and measuring 6–20 μm × 0.1 μm. One or both ends of the organism are hooked or recurved on the body. Requires dark-ground illumination for its demonstration if unstained (Fig. 87), silver impregnation stains are used to allow visualisation by normal illumination.

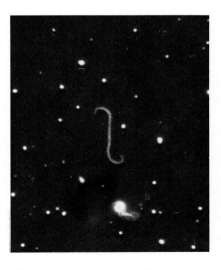

Fig. 87 Dark-ground preparation of *L. interrogans* serovar. *icterohaemorrhagiae* from Stuart's medium. Both ends of the organism are recurved on the body; the very fine coils are just discernible
× 1200

Cultural appearances

Aerobic. Extremely fastidious and requires special fluid media for growth, e.g. Stuart's medium. Intolerant of acid environments and optimal pH = 7.2–7.4. (N.B. Optimum temperature for *in vitro* growth is 30°C).

Serological characteristics

At present 18 serogroups are recognised on the basis of shared major agglutinogens and within each serogroup various serotypes can be identified by using selectively absorbed antisera; thus, for example, there are 13 serotypes within the icterohaemorrhagiae group including serovar *icterohaemorrhagiae, copenhageni,* etc. Another serogroup, i.e. canicola contains 11 serotypes, e.g. *canicola.*

Animal inoculation

Young guinea pigs and weanling golden hamsters are very susceptible to intraperitoneal inoculation either of cultures or of blood or urine specimens obtained from a patient. The animal dies in 1–2 weeks with jaundice apparent on serous membranes and haemorrhages in the lungs and muscles.

The ability of leptospirae to pass through skin is taken advantage of in attempting to demonstrate their presence in water; a young guinea-pig whose belly has been shaved is partially immersed in the water for 1 h. If infection takes place the animal, at post-mortem, displays the characteristics mentioned above.

SAPROPHYTIC LEPTOSPIRES

These are frequently present in water either in nature or from domestic taps. They are morphologically indistinguishable from pathogenic species but are readily differentiated from the latter in growing readily in simple media and their lack of pathogenicity to laboratory animals.

LEPTOSPIROSIS

Although specific names have been given to the clinical illnesses resulting from infection with leptospirae, e.g. canicola fever, Fort Bragg fever, seven-day fever, etc, the epidemiology of leptospiral infection is identical regardless of the particular serotype or natural host involved. The classical leptospiral infection is known as Weil's disease and the natural host of the causal leptospire, *L. interrogans* serovar *icterohaemorrhagiae,* is the brown rat *(Rattus norvegicus)* which frequently remains healthy. The leptospirae are excreted in the rat's urine.

Provided the urine is voided in moist, alkaline surroundings the spirochaetes may remain viable for many days and if man comes in contact with them they can penetrate skin and mucosal surfaces through cuts and abrasions.

Although infection may occur by bathing in polluted water most cases are associated with an occupational risk, i.e. occupations involving work in rat-infested places which are also moist, such as coal mines, fish-gutting halls or sewer systems. In similar fashion and in other countries workers in rice-fields or sugar-cane workers are at risk from other leptospires; different serotypes have adapted to different animal hosts and whilst these latter are usually rodents, *L. interrogans* serovar *canicola* has as its host the dog and also the pig.

Control of leptospiral infection depends on elimination of the natural host, the modification of the environment in occupations with a high risk, e.g. rodent proofing of fish-gutting halls and the spraying of working surfaces in such premises with an acid solution at the beginning and end of each working period, the provision of protective clothing for sewer workers, etc.; when cases of leptospirosis are traced to a domestic animal then the animal must either be sacrificed or treated to eliminate leptospiral parasitisation.

Direct transmission from man to man is extremely rare.

Legionellae

25

Members of this genus are probably saprophytes in water or soil from which sources man becomes infected.

LEGIONELLA PNEUMOPHILA

The generic name is derived from the first recorded outbreak of infection at a conference of American legionnaires in Philadelphia in 1976, and the specific epithet indicates a predilection for lung tissue.

Microscopy

Usually 1 μm $\times$ 2–3 μm but long forms up to 20 μm or greater occur. Gram-negative. Flagellate and fimbriate.

Cultural characteristics

Extremely fastidious; aerobic but growth enhanced by increased CO_2 tension (10%); L-cysteine required for growth, slow growing at 35–37°C and colonies may be noted at 3–5 days. Charcoal yeast-extract agar yields light blue-coloured colonies; on clear media colonies show brownish discoloration of the medium. At present the most sensitive isolation technique is to inoculate material on to the yolk sac of a fertile hen's egg; 7 days later the yolk sac contents are subcultured to solid media.

Biochemical activities

Apparently inactive against commonly used carbohydrates.

Serological characters

So far six serotypes of *L. pneumophila* have been recognised and detection of a significant rise ($>$ 4-fold) in antibody titre between acute and convalescent sera from a suspect case is, at present, the most common method of diagnosis.

To date six other species within the genus *Legionella* have been recognised but their differentiation is presently restricted to a few specialised laboratories.

LEGIONNAIRES' DISEASE

Although first recognised at a legionnaires' conference in the USA in 1976, the disease has no military connotation and is certainly not restricted to men; the infection has received publicity far above its importance either in morbidity or mortality and might equally well be called Reporters' or Mass-Media Disease, since in comparison to the more common forms of pneumonia it is rare and deaths are few. At present we must strive to keep this disease in perspective and remember that, e.g., in 1981 many more people *died* from pneumococcal pneumonia in the UK than the total number alleged to *suffer* legionnaires' disease.

Older age groups and heavy smokers seem to predominate as candidates but case-to-case spread is unusual and in the majority of outbreaks there is an association with humidifying plants and air-conditioning systems; however sporadic cases occur without such an association. *L. pneumophila* has been isolated also from spray-heads in shower cabinets, which reaffirms its primarily saprophytic nature.

The disease is not 'new'; surveys of human sera stored for other purposes have shown significant levels of antibody to *L. pneumophila* long before the disease or the organisms were recognised.

Since infection is more likely to be severe in older age groups and erythromycin therapy is effective, clinical suspicion of *L. pneumophila* infection combined with the radiological picture and recovery after erythromycin administration is often taken as proof of the aetiology without adequate serological or other investigation.

In addition to the original dramatic pneumonic illness, various other, and often milder, clinical presentations of infection have been encountered, e.g. of a mild influenzal nature with no mortality, or alternatively a dysenteric, short-lived, infection with minimal respiratory involvement.

Recent experimental evidence shows that *L. pneumophila* is pathogenic for several ubiquitous saprophytic amoebae living in fresh water and soil and it is probable that such parasitised amoebae, e.g. of the genus *Acanthamoeba,* may be the means of man being infected with the *Legionella* species.

A further spin-off for this finding is that the microbiologist might be able to use laboratory-maintained amoebae for the primary enrichment of *Legionella* species from water and other samples.

Actinomycetes

26

This heterogeneous collection of microorganisms may superficially resemble fungi but are related to true bacteria; characteristically they form a branching mycelium which tends to fragment into coccal and bacillary pieces. Many actinomycetes are entirely saprophytic, particularly in soil; a few produce disease in man and animals.

ACTINOMYCES

This genus comprises obligately anaerobic or microaerophilic members which are parasitic on man and some domestic animals; the two most important species are *Actinomyces israelii* (human in origin) and *Actino. bovis* from cattle and pigs, but they have many common biological characteristics.

Microscopy

Filamentous branching organism, Gram-positive, non-motile, non-capsulate and non-sporing. In tissues from infected individuals 'sulphur granules' (Fig. 88) are formed which comprise a central mycelial mass with a peripheral zone of swollen clubs which stain Gram-negatively and are acid-fast provided that only 1% H_2SO_4 is used in attempted decolorisation (Figs 89 and 90). Club

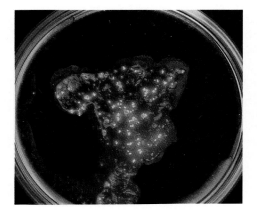

Fig. 88 'Sulphur granules'. These granules were obtained from a pleural effusion in a case of thoracic actinomycosis with secondary staphylococcal infection. The granules vary in size and a few of the largest granules are showing a darker, almost brownish colour

formation is more marked in bovine than in human lesions.

Cultural appearances

Primary isolation must be by anaerobic cultivation on blood agar or in a shake culture of serum agar in which colonies form in a band 10–20 mm below the surface of the medium. When grown as surface colonies there is much variation in appearance, but

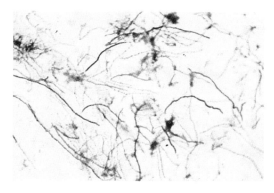

Fig. 89 Gram-stained film of *Actino. israelii* made from an anaerobic culture plate which had been inoculated with a sulphur granule
× 800

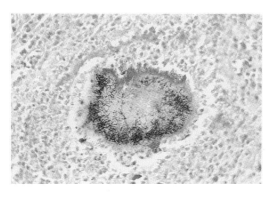

Fig. 90 Gram-stained section of a sulphur granule in tissue showing Gram-positive actinomyces mycelial growth surrounded by a peripheral zone of actinomyces 'clubs'
× 200

these usually present a rough appearance, irregular in outline, cream or white in colour and in the case of *Actino. israelii* are often firmly adherent to the medium (Fig. 89).

Biochemical abilities

Actino. bovis, unlike *Actino. israelii,* hydrolyses starch but otherwise their fermentative abilities vary with different strains within each species and do not allow reliable differentiation.

Serological characters

Four groups can be recognised but, at present, such techniques are not employed in diagnosis.

Animal inoculation

Results are irregular and of no value in diagnosis.

ACTINOMYCOSIS

Actinomycosis in the human is an endogenous infection with *Actino. israelii* derived from the mouth, where the actinomycetes live an essentially commensal existence: they can be isolated from tonsils and from carious teeth in 5% of healthy individuals.

The majority of cases of human actinomycosis occur in the cervico-oro-facial region (65°) with abdominal actinomycosis being next most common (19%); there is almost always a history of trauma preceding the onset of infection, e.g. extraction of a carious tooth or an accidental blow.

The disease presents many interesting epidemiological problems and, although the dissociation from bovine actinomycosis has been proven, male agricultural workers have a very much higher incidence of infection than men in other types of employment; similarly men suffer more frequently than women (3:1) and although the disease occurs at all ages, more than half the incidence occurs between the ages of 10 and 29 years. Case to case infection is unknown.

Recently species of *Actinomyces* have been found in association with certain intrauterine contraceptive devices and they may have a role in chronic pelvic inflammatory disease.

SECTION 3
Diagnostic methods

This section aims to give the student insight into the processing of specimens which are most commonly submitted to bacteriology laboratories, and cross reference to relevant chapters in Section II should be made when studying the various processing charts.

Advice regarding the collection and transmission of specimens is given in each of the following chapters but there are general points which must be borne in mind; these are apparently mundane but the constant trickle of specimens in which such elementary facts are ignored demands that they should be reiterated.

1. *Identification.* Patient's name, ward or address, age and sex on request form and suitable details on the specimen container.

2. *Clinical data.* A précis of the history or a statement of the patient's clinical diagnosis *including* what antimicrobial therapy (if any) has been initiated.

3. *Timing.* The date *and* time of the collection of the specimen must be stated.

In the processing charts throughout this section the following abbreviations are used to indicate commonly used culture media.

NA = nutrient agar
BA = blood agar
CVBA = crystal violet blood agar
McC = MacConkey's medium
DCA = deoxycholate citrate agar
CMB = cooked meat broth
W&H = Willis & Hobbs's medium

Blood culture

27

Bacteraemia is a constant finding in many human infections and usually occurs in the early stages of illness; the attempted isolation of organisms from the blood is undertaken less frequently than it might be since the identification of an accepted pathogen from the blood stream offers a definite and probably early diagnosis; e.g. in the enteric fevers the causal salmonellae can be isolated from the blood very readily during the first week of the illness when isolations from the stool are less common.

Materials required at bedside (Fig. 91)

(a) Soap and water, surgical spirit and povidone-iodine with swabs to apply the cleansing agents.
(b) Tourniquet to render veins turgid in the ante-cubital fossa.

(c) 10 ml sterile disposable syringe and needle with a second sterile needle.
(d) Blood culture bottles; these contain 50 ml of one of several media and frequently incorporate 0.05% sodium polyanethyl sulphonate (Liquoid) which acts as an anticoagulant and also annuls the natural bactericidal action of blood.

The screw-cap top of the bottle is kept sterile by a 'viskap' or another suitable cover which can be removed by pulling upwards on the linen or metal rip-cord to expose the sterile metal cap with a central hole under which is a rubber diaphragm through which the blood is injected from the syringe (Fig. 92). By this method the contents of the blood culture bottle are not exposed to the air during inoculation.

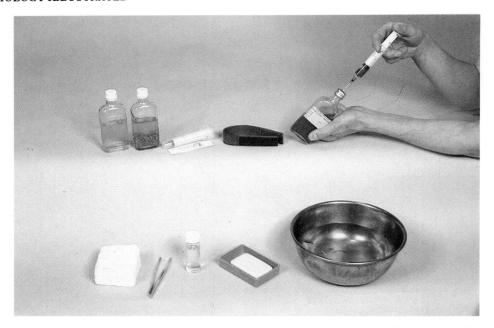

Fig. 91 On the top row are two blood culture bottles, that on the left comprising brain-heart infusion broth and its immediate neighbour is a home-made cooked meat broth; a sterile disposable syringe and needle and a tourniquet. The method of inoculation is shown at the upper right.

Cleansing agents, namely gauze swabs, surgical spirit and water, are shown in the bottom row

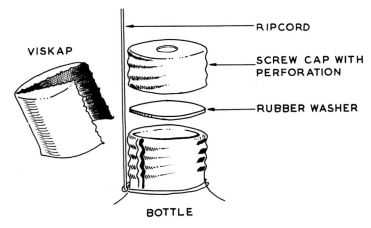

Fig. 92 Exploded view of top of blood culture bottle

Bedside procedure

An assistant should be available to control the tourniquet, rip off the 'viskap' and attend to the patient's arm after venepuncture.

The ante-cubital area should be thoroughly washed with soap and water, then with surgical spirit and finally treated with povidone-iodine, in order to reduce as far as possible contamination of the specimen with skin organisms. The tourniquet is applied and if necessary the veins made more prominent by the patient clenching and opening his hand; the syringe is removed from its sterile container without touching the needle and 5 to 10 ml of blood are withdrawn from a vein. The tourniquet is then released, the syringe and needle withdrawn and pressure applied to the puncture area.

The needle is detached from the syringe and, using forceps, is replaced by a fresh sterile needle; the 'viskap' is ripped off the blood culture bottle, the rubber diaphragm is pierced with the freshly attached sterile needle and the contents of the syringe are injected into the bottle, which is then agitated gently for a few seconds to ensure that the blood does not clot.

The syringe and needle are withdrawn from the diaphragm as one unit and the rubber diaphragm is swabbed with surgical spirit.

The use of the second needle to inoculate the blood culture bottle eliminates the prospect of isolating skin-contaminating bacteria gathered on the needle used for venepuncture.

Laboratory procedure

The inoculated bottles are returned immediately to the laboratory, and most laboratories provide a free-access incubator for the receipt of such bottles outside normal working hours. Incubation is normally at 37°C, and after 18–24 h two or three loopsful of the blood-broth mixture are removed aseptically and inoculated on to two blood agar plates, one of which is then incubated aerobically and the other under anaerobic conditions.

This procedure is repeated frequently for at least 2–3 weeks.

Various media are available for blood culture but *as a routine* two bottles should be inoculated from any one blood specimen. One of these should be a cooked meat broth medium and recent extensive investigations in several centres have shown that such cooked meat medium is best made in the laboratory since the home-made product gives superior

isolation of fastidious organisms in comparison with the mass-produced commercial equivalent. One reasons for this superiority is that cooked meat broth has a restricted shelf life and the time lapse between mass-produced medium and its eventual use exceeds that shelf life; laboratories producing their own media have a 'time-expired' date stamped on the label. Unused time-expired cooked meat blood culture bottles should be returned to the laboratory where the contents can be rejuvenated.

Numerous special blood culture media can be obtained when the clinician suspects that the bacteraemia is due to a particular bacterium, e.g. liver infusion broth for suspect acute brucellosis, saponin broth in cases of suspect *Strept. viridans* endocarditis.

All blood culture bottles contain suitable concentrations of p-amino-benzoic acid and β-lactamase to nullify respectively the activity of sulphonamides and penicillins should these have been used in treating the patient.

CLOT CULTURE

In the case of the enteric fevers clot culture yields results quantitatively superior to the blood culture technique described above.

For clot culture 5 ml of venous blood is collected aseptically in a screw-capped bottle and allowed to clot; the separated serum is then aseptically removed and 15 ml of 0.5% bile-salt broth containing 100 units/ml of streptokinase is added to the clot. Rapid lysis of the clot occurs and the mixture is then examined as for blood culture.

The advantages of clot culture are that the clinician does not require a supply of special blood culture bottles, that isolation of salmonellae is superior to normal blood culture techniques and that a Widal test can be performed on the serum, thus giving a basal titre for the particular patient against which subsequent antibody determinations may be judged.

Urine

28

SPECIMEN COLLECTION

Catheterisation is to be avoided whenever possible since it carries a definite risk of *causing* infection, either through faulty aseptic techniques or by contamination with bacteria residing in the urethra. Certainly catheterisation is never indicated solely to provide a specimen of urine for bacteriological testing; properly collected mid-stream specimens are entirely suitable in both sexes.

In males the prepuce should be retracted and the glans washed thoroughly with soap and water.

In females the patient is instructed firstly to wash and dry her hands and then using sequentially four separate sponges soaked in 10% green soap solution to wash the vulva thoroughly working from the front backwards; the labia are held apart with the fingers of one hand whilst the specimen is being passed.

The patient is given a sterile wide-mouthed container, e.g. a 1 lb (2.2 kg) honey pot, and instructed so that after micturition has commenced the container is inserted into the established flow of urine without interrupting micturition.

Since many bacteria flourish in urine (Pasteur used urine as a culture medium) it is essential that the specimen be examined within 3 h of collection, i.e. before organisms which may be present enter the logarithmic growth phase, otherwise artificially high counts will result.

If examination must be delayed beyond this 3 h period the specimen should be maintained at 4°C until it reaches the laboratory.

LABORATORY PROCEDURE (Fig. 93)

Bacterial count

Dilutions of 1 in 100 and 1 in 1000 are made from the uncentrifuged specimen and incorporated in pour plates or spread over the surface of nutrient agar plates; after overnight incubation the resultant colonies are counted and the number of bacteria per ml of specimen estimated. Counts of 10^3 organisms or less per ml are indicative of contamination whereas counts of 10^5 organisms or more per ml are associated with infection in the urinary tract. Specimens yielding 10^4 organisms per ml may be difficult to interpret, but where several bacterial species are present such counts are most likely to mirror contamination and not infection; when counts in this region are made, further specimens should be examined to assist evaluation in a particular case.

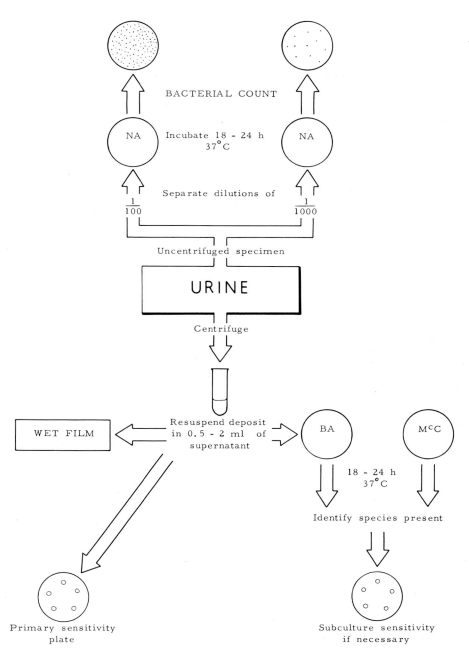

BACTERIAL COUNT

NA   Incubate 18 - 24 h   NA
            37°C

Separate dilutions of

$\frac{1}{100}$ $\frac{1}{1000}$

Uncentrifuged specimen

URINE

Centrifuge

WET FILM ⟵ Resuspend deposit ⟶ BA   M^cC
              in 0.5 - 2 ml of
              supernatant

                              18 - 24 h
                                37°C

                        Identify species present

Primary sensitivity   Subculture sensitivity
plate   if necessary

Fig. 93 Laboratory examination of urine sample

It is impracticable in a busy laboratory to submit all specimens to this method of enumeration, and screening techniques are in use; these may be carried out using commercial materials which comprise a small plastic plate, one side of which is coated in nutrient agar and the other with MacConkey's medium. This coated plate is dipped into the well-mixed specimen of urine and then placed in a sterile screw-capped container and incubated. Subsequent growth can be enumerated against standard growth charts; the so-called dip slide method

An even simpler method uses standard blotting paper strips with one end bent at right angles; this foot is dipped into the specimen and then withdrawn and the strip held with the foot upwards so that excess urine can be absorbed into the limb of the blotting paper before the foot area is impressed on the surface of a nutrient agar plate. A second strip is used to inoculate an area on a plate of MacConkey's medium.

The number of colonies present in the inoculated area can be equated with the bacterial population per ml of the specimen by comparison with standard graphs.

Cultivation

5 to 10 ml of the specimen is centrifuged at 3000 r/min for 5 min and the deposit is resuspended in a suitable volume of supernatant. A wet film is made and examined microscopically for the presence of bacteria, pus cells and red blood cells; if this examination shows any bacteria a primary sensitivity plate is seeded from the deposit

and discs of sulphonamide, trimethoprim, cephalosporins, etc are applied. Finally, a blood agar and a MacConkey plate are inoculated with the deposit and incubated overnight. Following species identification, subculture sensitivity tests may, if necessary, be performed from the isolated colonies on the diagnostic media.

In the majority of cases of urinary tract infection only a single species is involved and the most commonly occurring are members of the enterobacteria, particularly *Esch. coli;* when two or more species are isolated then further freshly-voided specimens should be examined. Only if the same species are repeatedly found together might one assume that both are involved in the infection.

The recent finding that a significant proportion of cases of cystitis in young females are caused by coagulase-negative staphylococci must be remembered; likewise there is growing evidence that lactobacilli and bacteroides species may be causes of urinary tract infection. The ultimate proof that a particular species is causing infection can be obtained by suprapubic aspiration of urine from the full bladder since, with proper skin preparation, such specimens cannot be contaminated from the distal urethra, vulva or vagina.

Cerebrospinal fluid

29

SPECIMEN COLLECTION

Iatrogenic meningitis still features unfortunately as a result of faulty technique during lumbar puncture and it is essential that full surgical aseptic technique should be used to collect the specimen. If the CSF cannot be delivered to the laboratory *immediately*, the container and its contents should be maintained at 37°C; in any case laboratory procedures should be undertaken within 2–4 h of the specimen being obtained.

LABORATORY PROCEDURE (Fig. 94)

In acute pyogenic infections the fluid is usually turbid in appearance but can vary from a clear to a grossly purulent state — a clear fluid is often seen in cases of aseptic meningitis.

One ml of uncentrifuged CSF should be removed aseptically and added to a tube containing 1 ml of 0.2% glucose broth. This favours the growth of meningococci and pneumococci without disturbing the growth of *H. influenzae* or other, rarer, causal organisms. After incubation at 37°C for 18 to 24 h, subculture is made to blood agar plates which are incubated similarly to the

procedure outlined in the diagram.

A cell count should be performed, and then the remaining fluid is centrifuged at 3000 r/min for 5 min and 2 ml of supernatant fluid is transferred to a sterile container and submitted for biochemical examination, i.e. the estimation of the glucose, protein and chloride content. The deposit is seeded on to two blood agar plates one of which is incubated aerobically and the other in an atmosphere of air plus 5% CO_2. A Gram film is made from the deposit and the remaining fluid incubated overnight.

N. meningitidis appears as intracellular Gram-negative diplococci which may vary in number; pneumococci and *H. influenzae* also occur in clinically identical cases. Staphylococci and β-haemolytic streptococci are encountered in occasional cases, and indeed almost all human pathogens have been isolated from CSF. The identification of the three commonly occurring organisms mentioned above must be fulfilled by cultural and biochemical methods and tested for sensitivity to antibiotic agents before a final report is issued, but a tentative report can be made on the basis of the Gram-stained preparation.

The potential lethality of acute bacterial meningitis regardless of the age or sex of the patient has stimulated the search for faster

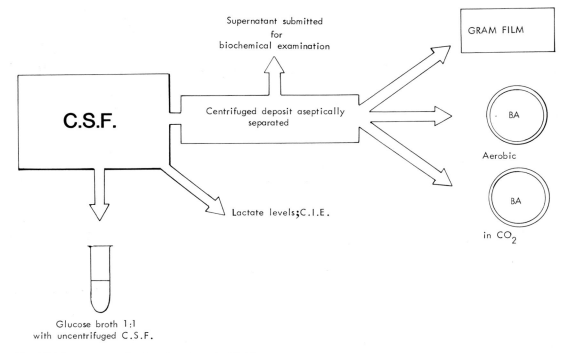

Supernatant submitted
for
biochemical examination

GRAM FILM

C.S.F.

Centrifuged deposit aseptically
separated

BA

Aerobic

BA

in CO_2

Lactate levels; C.I.E.

Glucose broth 1:1
with uncentrifuged C.S.F.

Fig. 94 Laboratory examination of cerebrospinal fluid

methods of diagnosis, and the determination of lactic acid levels by gas-liquid chromatography is one such; levels of less than 35 mg/dl virtually exclude acute bacterial meningitis, although such levels are seen in viral meningitis. Results are available within 30 min.

Similarly, using counter current immunoelectrophoresis (C.I.E.) the patient's CSF can be assayed for the presence of bacterial antigen using standard sera against *N. meningitidis*, pneumococci and *H. influenzae*; results can usually be read within 60–90 min.

Concomitant blood cultures should be submitted, since in many instances a bacteraemia precedes or is coincident with the meningeal illness.

Throat Swabs

30

SPECIMEN COLLECTION

The throat should be examined with adequate illumination and a spatula must be employed so that the affected area can be swabbed accurately and without contamination from the buccal cavity; swabs should not be taken within 6 h of gargling or of the administration of antimicrobial agents. If more than 12 h delay is likely before the swab reaches the laboratory then serum-coated swabs are advantageous in ensuring the survival of *Strept. pyogenes.*

Saliva is an acceptable alternative or additional specimen in suspect cases of streptococcal sore throat and is the specimen of choice when the dental practitioner wishes knowledge of the buccal flora before operating on a patient who, because of endocardial defects, requires peroperative antimicrobial cover to reduce the risk of subacute bacterial endocarditis.

LABORATORY PROCEDURE (Fig. 95)

Strept. pyogenes, Vincent's organisms and *C. diphtheriae* must be sought in all specimens; in individuals being screened before dental surgery the commensal flora must also be

evaluated and the antibiogram of any viridans-type streptococci tested and reported.

STREPT. PYOGENES

Crystal violet blood agar (CVBA) plates are inoculated; the incorporation of a 1 in 500 000 concentration of crystal violet reduces the growth of commensal organisms. A bacitracin-impregnated disc is placed in the well-inoculum of each plate; strains belonging to groups A, C & G are inhibited whilst those belonging to other groups grow freely in the immediate vicinity of the bacitracin disc. One of the plates is incubated anaerobically since this favours the growth of β-haemolytic streptococci in comparison with the rest of the throat flora; colonies of *Strept. pyogenes* are larger and show wider zones of β-haemolysis under anaerobic conditions.

C. DIPHTHERIAE

A Loeffler slope and a plate of tellurite medium are inoculated; growth from the Loeffler slope is rapid and films can be made after 12 to 18 h incubation and stained by Gram's and Albert's methods. Volutin

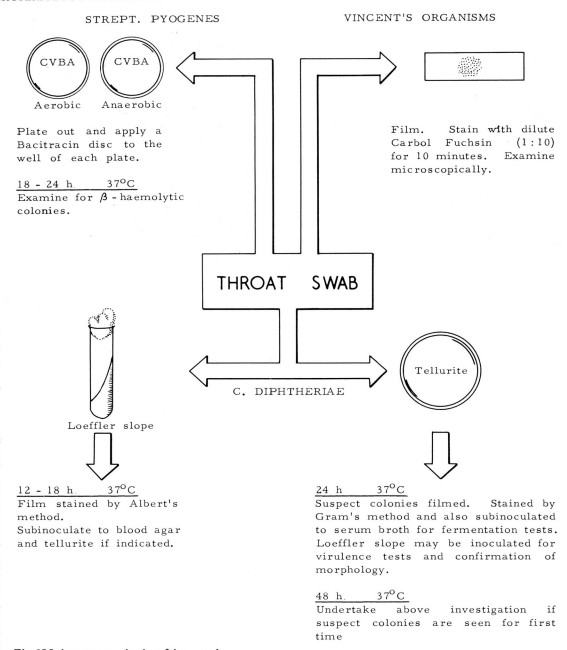

STREPT. PYOGENES

VINCENT'S ORGANISMS

Plate out and apply a
Bacitracin disc to the
well of each plate.

18 - 24 h   37°C
Examine for β - haemolytic
colonies.

Film.   Stain with dilute
Carbol Fuchsin   (1:10)
for 10 minutes.   Examine
microscopically.

THROAT   SWAB

Loeffler slope

C. DIPHTHERIAE

Tellurite

12 - 18 h   37°C
Film stained by Albert's
method.
Subinoculate to blood agar
and tellurite if indicated.

24 h   37°C
Suspect colonies filmed.   Stained by
Gram's method and also subinoculated
to serum broth for fermentation tests.
Loeffler slope may be inoculated for
virulence tests and confirmation of
morphology.

48 h.   37°C
Undertake   above   investigation   if
suspect colonies are seen for first
time

Fig. 95 Laboratory examination of throat swab

granules develop rapidly on this medium and
the detection of Gram-positive bacilli which,
with Albert's stain, show volutin granules
demands subinoculation to blood agar and
tellurite media so that colonies can be

obtained in a pure state for further
identification. Colonies on tellurite media
may not be recognisable for 48 h; if any
colonies are regarded as those of *C.
diphtheriae* they should be subcultured after

their morphology has been studied. All suspect colonies must be subjected to fermentation tests and if these confirm their identity as *C. diphtheriae*, the final proof of their virulence is by demonstrating toxigenicity.

Remember that although bacteriological confirmation of the diagnosis must be sought *treatment must not be deferred* until bacteriological results are forthcoming.

VINCENT'S ORGANISMS

A film is made from the throat swab and stained with dilute carbol fuchsin (1 in 10) for 10 min. The presence of *large numbers of Borr. vincentii* and fusiform bacilli is indicative of Vincent's infection.

With the above exception, stained films made directly from throat swabs are of no value in the diagnosis of throat infections since commensal cocci and bacilli will be found which morphologically are indistinguishable from *Strept. pyogenes* and *C. diphtheriae*. The swab should be used to inoculate the Loeffler slope, CVBA plates and the tellurite medium in that order so that the relatively inhibitory substances in the plate media are not carried over to the Loeffler slope. The film for Vincent's organisms is made last of all.

Sputum

31

SPECIMEN COLLECTION

One of the most frequent frustrations in diagnostic laboratories is to receive a specimen of 'sputum' which is entirely salivary in origin. It is essential that the specimen should be coughed up and not merely result from expectoration of saliva or hawked from the post-nasal space. Many of the pathogenic species involved in chest infections are feebly viable outside the host's tissues so that the specimen should be delivered to the laboratory rapidly.

The collaboration of the physiotherapist frequently ensures that a proper specimen is obtained, particularly when the patient has a dry cough.

Transtracheal aspiration, although rarely practised in this country, ensures that the specimen is not contaminated from the oro-pharynx.

LABORATORY PROCEDURE (Fig. 96)

Since all respiratory tract pathogens, excepting tubercle bacilli, can be isolated (usually in small numbers) from the healthy upper respiratory tract, and since by direct sampling of the raw specimen in different parts different organisms may be cultured, it is essential that the specimen be homogenised before plating; thus a clearer picture of the relative proportions of different bacteria can be obtained.

Various methods of homogenisation are employed, from simple mechanical agitation of the specimen after the addition of sterile glass beads to various chemical techniques, e.g. an equal volume of 1% buffered pancreatin is added to the sputum and the container thoroughly shaken, it is then placed in a 37°C water bath and shaken every 15 min for 1 h. Thereafter a Gram-stained film is made and examined microscopically for cellular exudate and bacteria.

Laboratory practice varies in regard to the media routinely inoculated with sputum specimens; the minimal requirement is cultivation on a blood agar plate incubated aerobically. In the well-inoculum are placed an Optochin disc for rapid differentiation of *Strept. viridans* and pneumococci and a penicillin disc which will inhibit the growth of many other organisms whilst allowing *H. influenzae* to flourish. A primary sensitivity plate should be inoculated in parallel with the diagnostic medium. Preferably a second blood agar plate should be inoculated and incubated in an atmosphere with added CO_2 (5%) since occasional strains of pneumococci and *H.*

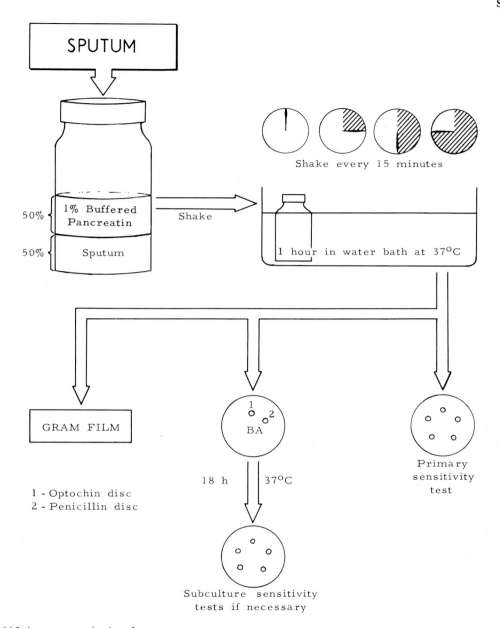

Fig. 96 Laboratory examination of sputum

influenzae will not grow on primary isolation except in such an atmosphere.

The processing of specimens of sputum from suspect cases of pulmonary tuberculosis is dealt with in a later chapter but it should be noted that in such cases at least six consecutive daily specimens should be examined.

Faeces

32

SPECIMEN COLLECTION

Whenever possible the specimen should be examined within a few hours of being passed, and if more than 12 h delay is likely before reaching the laboratory a suitable preservative should be added — e.g. an equal volume of buffered glycerol saline solution — and the entire contents of the container should be thoroughly mixed.

Rectal swabs are not infrequently submitted but these are less satisfactory than stool specimens even when the swab has been properly used, i.e. by passing beyond the internal sphincter; pathogenic species survive less readily on swabs and equally the bacteriologist is denied the examination of material microscopically, e.g. for cellular exudate.

The range of potentially pathogenic species sought by the laboratory must include campylobacters, salmonellae and shigellae, since the first of these now rivals members of the others as the commonest bacterial pathogen; in addition, laboratories should examine faecal specimens for enteropathogenic *Esch. coli* and, if indicated on epidemiological grounds, *V. parahaemolyticus.* Search for less frequently encountered pathogens or their toxins, e.g. *Cl. difficile* is not routinely performed.

LABORATORY PROCEDURE (Fig. 97)

Any specimen which is at all formed must be thoroughly emulsified in sterile physiological saline before being plated on to diagnostic media or used to inoculate fluid enrichment media.

CAMPYLOBACTER JEJUNI

The unusual conditions for optimal growth must be catered for, i.e. incubation temperature for media is 43°C and the environment must be microaerophilic; the latter is most readily obtained by using gases from a cylinder in the proportion of 5% oxygen, 10% CO_2 and 85% H_2 and incubating inoculated media in an anaerobic jar after evacuation of the normal atmosphere and replacement with the mixed gases.

Several selective media are employed for the isolation of *C. jejuni* and depend primarily on the incorporation of antibiotics, e.g. a combination of trimethoprim, vancomycin and polymyxin B.

SHIGELLAE AND SALMONELLAE

The specimen is plated out on selective media (DCA and MacConkey's) and also inoculated

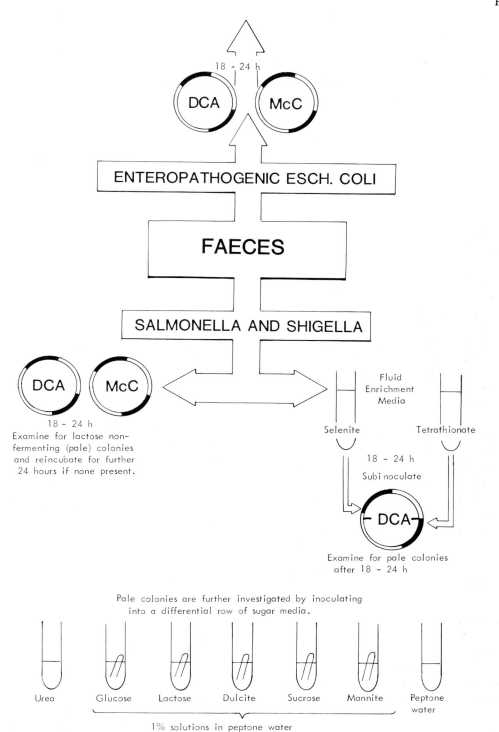

Fig. 97 Laboratory examination of faeces

into tubes of selenite and tetrathionate enrichment broths. After overnight incubation these broth cultures are subinoculated to a fresh DCA plate. On either of the selective media, lactose non-fermenting (pale) colonies must be regarded as shigellae or salmonellae until proven otherwise; several such colonies from each plate should be submitted to biochemical tests. The inclusion of a tube of urea allows rapid elimination of urease producers which are not intestinal pathogens. The final type identification of a member of either pathogenic genus is by serological methods.

ENTEROPATHOGENIC *Esch. coli*

In addition to inoculating a MacConkey and DCA plate (strains grow poorly if at all on the latter medium), a blood agar plate should be inoculated lightly since occasional strains grow only on this medium. There are no enrichment media for *Esch. coli*. On MacConkey's or BA media, colonies of enteropathogenic strains are identical with all other types of *Esch. coli.*, nor is there any method of biochemical differentiation; therefore it is necessary to test, by slide-agglutination, at least 10 colonies from each plate with a polyvalent antiserum prepared against the six enteropathogenic types. If agglutination is noted then the reacting colony is further tested against each of the individual type-specific antisera.

It should be appreciated that the macroscopic characters of faeces in cases of Sonne dysentery are only rarely of the classical type, i.e. a mixture of mucus and blood; usually the stool is unaltered apart from its consistency. Whilst wet film preparations may reveal the presence of pus cells and/or red blood cells, Gram-stained preparations of faeces are of no assistance in diagnosis on account of the morphological similarity of pathogens and the commensal coliform flora.

Pus and wound exudates

33

SPECIMEN COLLECTION

There are at least two reasons why a volume of pus or part of an excised wound should be submitted for examination rather than a swab sample; firstly, the swab specimen will dry out to a greater or lesser extent and many pathogenic bacteria, e.g. *Bacteroides,* will not survive. Similarly the number of media employed in examining such specimens implies a reducing inoculum over several plates and tubes of culture media. If swabbing is the only practical method of sampling then at least three swabs must be submitted, one swab being used to make films for staining and microscopic examination and the others for inoculating culture media. The inner dressing from an infected wound, placed in a sterile honey jar, is another alternative.

Speed of transmission is vital since anaerobic species do not tolerate exposure to the normal atmosphere nor to desiccation; ideally specimens should be collected by a member of the laboratory team and inoculated on to media at the bedside.

LABORATORY PROCEDURE (Fig. 98)

1. Macroscopic examination should always

be made if actinomycosis is a possible diagnosis. A particular search should be made for sulphur granules which may in early cases of such infection be atypical, i.e. white in colour or even semi-transparent.

2. Microscopic examination always requires Gram-stained films and, again, if granules are seen with the naked eye one should be removed and crushed between two slides before staining. In cases of suspected tuberculous infection a Ziehl-Neelsen film should be made directly from the material and part of the specimen treated as outlined in the next chapter.

3. Cultural methods must, as a routine, include the attempted isolation of aerobic pyogenic organisms, e.g. *Staph. pyogenes* and *Strept. pyogenes,* and also a search for anaerobic species, e.g. members of the genus *Bacteroides.*

Recovery of the popular pyogenic organisms, of which more than one species may be present in the specimen, requires the inoculation of two blood agar plates, one incubated aerobically and the other under anaerobic conditions; a MacConkey agar plate should also be inoculated. Additional cultures may be made on other media depending on the microscopic findings and/or the clinical history; in any case a primary sensitivity plate should be set up.

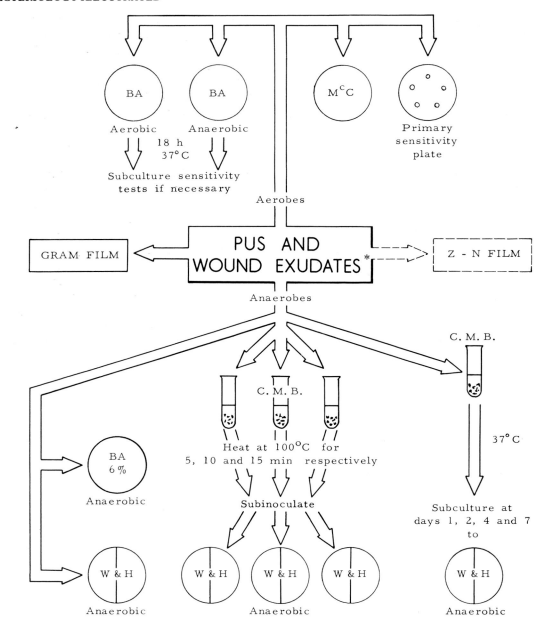

Fig. 98 Laboratory examination of pus and wound exudates

Until recent years the attempted isolation of anaerobic species centred classically on members of the genus *Clostridium*; since such organisms may occur in a wound solely as contaminants without producing infection, the diagnosis of tetanus or gas gangrene must be made on clinical grounds and treatment instituted without awaiting bacteriological confirmation.

A plate of blood agar (agar content increased to 6% to reduce the spreading tendency of *Clostridia*) and a plate of half-antitoxin Willis and Hobbs's (W & H) medium are inoculated directly from the specimen. The half-antitoxin W & H plate, using a mixture of *Cl. perfringens* type A and *Cl. oedematiens* type A antitoxic sera smeared over one half of the plate, is specially valuable in identifying *Cl. perfringens* and *Cl. oedematiens*.

More recently the dominance of *Bacteroides* species in infected wounds, particularly following colonic surgery, appendicectomy and hysterectomy, has had an impact on laboratory methods unrivalled for several decades; the need for adequate specimen taking and speedy transmission for processing has been mentioned already.

The isolation of *B. fragilis* is enhanced by using blood agar plates incorporating an aminoglycoside such as gentamicin and after inoculation the plates should be incubated under anaerobic conditions incorporating a 10% atmosphere of CO_2 for 48 h. Additionally tubes of cooked meat medium enriched with yeast extract should be inoculated and after 24 h incubation subcultures made to blood agar media. These again are incubated anaerobically. Primary BA plates and those subinoculated from cooked meat culture media should be inspected briefly at 24 h and any apparent growth subcultured and colonies filmed and stained by Gram's method; plates should be incubated for a further 24 h.

At present the constant sensitivity of *Bacteroides* species to metronidazole can be assumed but a check should be made of such sensitivity to detect the possible emergency of strains resistant to this antibiotic.

Tuberculous infection

34

SPECIMEN COLLECTION

Tuberculosis can affect virtually every tissue in the body, thus the specimen submitted may vary widely. Whilst there should be no unnecessary delay in forwarding specimens for examination, the hardiness of tubercle bacilli allows them to survive very readily. Cerebrospinal fluid is prevented from clotting by adding sterile sodium citrate; serous fluids, e.g. pleural effusion, are similarly treated. When urine is to be submitted for examination the specimen may be either a 24 h collection or three pooled, consecutive, early-morning specimens; cleansing of the genitalia with soap and water will reduce the population of *M. smegmatis*.

LABORATORY PROCEDURE (Fig. 99)

Direct films from the specimen, stained by the Ziehl-Neelsen method, are valuable only if the material contains very large numbers of bacilli; alternatively films may be stained with auramine and this allows the survey of a film to be carried out speedily and at a lower magnification. This latter method is particularly useful in countries which still have a heavy burden of tuberculous infection and where many films from suspect cases of infection have to be scanned.

All specimens which are not fluid and/or those which may contain other organisms must be subjected to concentration before further processing; numerous concentration techniques are available but all have the same aims, i.e. to liquefy the specimen thus releasing any tubercle bacilli, to destroy all other bacteria and, by centrifugation, to harvest any *M. tuberculosis* in a small volume. One concentration method is a modification of Petroff's technique: a volume of the specimen is mixed with an equal volume of 4% caustic soda and thoroughly shaken; the mixture is incubated at 37°C for 30 min and the container shaken thoroughly every 5 min. After incubation and shaking the mixture is centrifuged for 30 min at 3000 r/min and the supernatant fluid is discarded; a drop of phenol red solution is added to the deposit which is then very carefully neutralised with 8% HCl.

The concentrate is used to make films to be stained by the Z–N method, and to inoculate slopes of Lowenstein-Jensen (L–J) media; cultivation must always be undertaken since it is much more reliable than microscopy and also the bacilli are then available for sensitivity testing. Guinea-pigs may also be inoculated with the concentrate; 0.5 ml of the

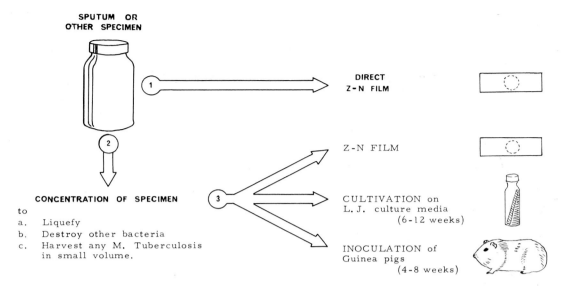

Fig. 99 Isolation of *Mycobacterium tuberculosis*

concentrate is injected into the left thigh muscles of an adult guinea-pig. Inoculation yields as high a proportion of positive results as does cultivation; with certain specimens, e.g. CSF and urine, inoculation yields a higher proportion of positive results than cultivation of the same material. Because of the risk of death from intercurrent infection it is usual to inoculate two guinea-pigs with material from each specimen.

The enteric fevers

35

These include typhoid and the paratyphoid fevers which in their essential clinical features are exactly alike; similarly their gross pathology is the same so that only by isolating and specifically identifying the causal organism can a distinction be made. *S. paratyphi B* is the commonest organism encountered in Britain and cases infected with *S. typhi* have usually acquired the infection overseas; *S. paratyphi A* and *S. paratyphi C* are rarely found in Britain.

The progression of clinical events illustrated is that of an untreated case; specific therapy, e.g. with chloramphenicol, rapidly eliminates bacteraemia, reduces the duration of pyrexia and alleviates rapidly the toxaemic symptoms.

The chart (Fig. 100) indicates the time of illness when isolation from the bloodstream and the faeces is most likely to succeed; microscopic examination of these specimens has no place in diagnosis.

Blood culture is performed as in Chapter 27 and the clinician should obtain from the laboratory blood culture bottles containing 0.5% sodium taurocholate broth, which medium enhances the isolation of enteric fever bacilli. Blood culture is most likely to be positive during the first 10 days of clinical illness and during relapses; three specimens should be obtained at 6–12 h intervals. It

should be noted that within 3 h of chloramphenicol therapy being instituted blood cultures are most unlikely to yield positive results.

Clot culture has also been mentioned in Chapter 27 and the advantages of this method merit its wider use; *bone marrow culture* is as reliable as blood culture and remains positive for 1–2 days after chloramphenicol therapy has begun but it is doubtful if such a procedure could be justified except in a minority of cases.

Faeces culture. The causal organisms can be isolated from the stool throughout the course of the illness but are most readily and frequently isolated during the second and third weeks; it is important that repeated examination of the faeces is undertaken since this increases the proportion of positive isolations.

Bile culture. Bile, aspirated by means of a duodenal tube, may be cultured in an attempt to isolate the causative organisms; this method is perhaps best restricted to carrier detection.

Urine culture. Bacilluria is intermittent in enteric fever cases and if attempts at isolation from urine are to be successful then daily examination of the morning urine for one week should be made. Even then, urine culture is not so frequently positive as stool culture; it is most likely to yield positive

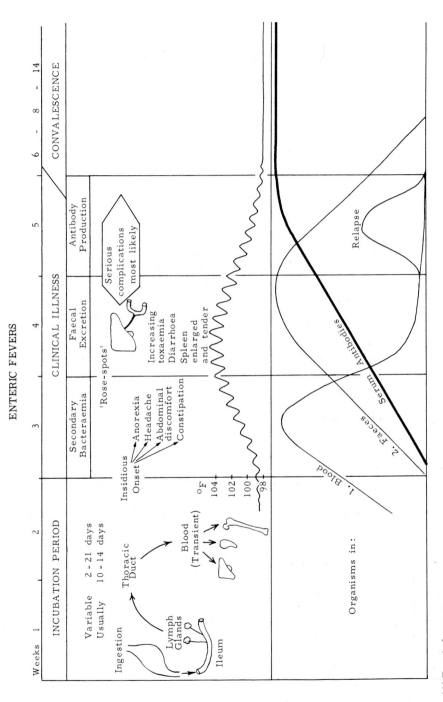

Fig. 100 Enteric fevers

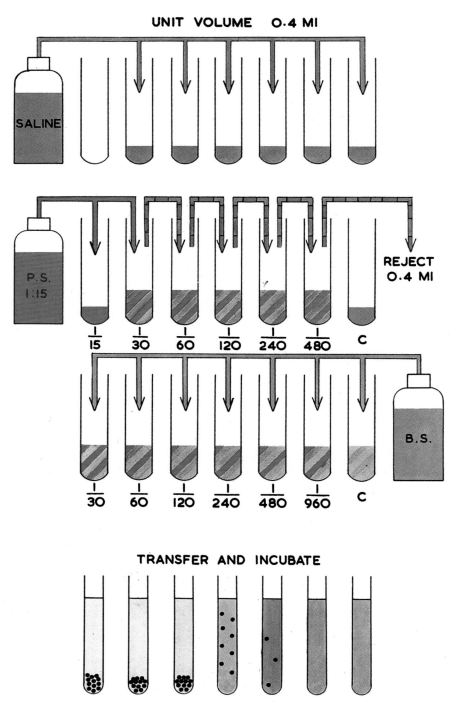

Fig. 101 *(see opposite)*

results after the second week of clinical illness. In the detection of carriers and determining whether carriage is intestinal and/or renal, care must be taken to ensure that stool specimens are not contaminated by urine, since if the individual is solely a urinary carrier such an error might lead him to be regarded also as a faecal carrier.

Serological tests

The Widal agglutination test (Fig. 101) is historically linked with the enteric fevers but the technique is also applied in other fields. In the enteric fevers the reaction is usually positive by the 7th to 10th day and the titre reaches its maximum during the 4th week. The patient's serum should be tested against both O and H suspensions of each of the relevant serotypes since occasionally only O or H agglutinins are detectable, especially in the early stages of the infection.

Interpretation of Widal test. Depending on the country in which an individual lives, and even on the community within any country, his serum may agglutinate the test suspensions in low dilution and such reactions are without diagnostic significance. In Britain the usual limits of such normal reactions are, for *S. typhi* and *S. paratyphi B*, H agglutination = 1 in 30, O agglutination = 1 in 50 and for the two other serotypes both O and H reactions = 1 in 10. In a case of enteric fever such low titres will be commonplace in the first week of the clinical illness so that a second serum sample must be tested 7–10 days later and will show a significant increase in titre. Anyone inoculated with typhoid vaccine also possesses specific agglutinins and these may complicate the interpretation of the Widal reaction; in such individuals a definitely rising titre for any one of the organisms might be regarded as diagnostically significant. Other non-enteric infections may cause an increase in titre which, however, falls rapidly on recovery unlike the maintenance of titre in enteric fever. The clinician should inform the bacteriologist of a history of typhoid vaccination.

Fig. 101 Quantitative agglutination test. Quantitative testing of serum for agglutinins towards bacteria is most frequently employed in suspect cases of the enteric fevers and contacts of such cases—the Widal test. Such tests can be applied in brucellosis and several other infections provided suitable standardised bacterial suspensions are available.

The patient's serum (P.S.) is initially diluted 1 in 15 with sterile physiological saline.

0·4 ml of saline is pipetted into tubes 2–7 and then 0·4 ml of 1 in 15 P.S. is pipetted into each of tubes 1 and 2, so that the serum dilution in tube 2 is now 1 in 30; serial dilution is now effected by thoroughly mixing the contents of tube 2 and then transferring 0·4 ml into the unit volume of saline in tube 3 and so on up to and including tube 6, from which, after mixing the contents, 0·4 ml is rejected. The dilutions now read from 1 in 15 through to 1 in 480 and the seventh tube, without serum, acts as a control in which only the bacillary suspension is added to ensure that the latter suspension is not auto-agglutinable. Finally, 0·4 ml of the standard bacillary suspension is added to each tube in the series, using a fresh sterile pipette and working from tube 7 through to tube 1 so that if any serum dilution is carried over from one tube to another the effect on the dilution is less than if the carry-over were from the lower dilution to the next highest.

Finally, the contents of each tube are thoroughly mixed and then transferred by individual capillary pipettes to agglutination tubes which are then incubated in a water bath at 37°C for 4 h.

Agglutination is indicated by clearing of the supernatant fluid to an extent depending on the amount of bacterial suspension which has deposited.

The titre of a particular serum is indicated by the tube containing the highest dilution of serum in which agglutination is noted by the naked eye

Leptospirosis

36

Figure 102 outlines the clinical phenomena at various stages of Weil's disease but it must be noted that the disease varies in severity and also that inapparent infections may occur in individuals exposed occupationally to the risk of infection. Furthermore, leptospirosis caused by the 30 or more other members of the genus is usually much milder than Weil's disease and has a correspondingly lower case fatality rate. The Figure also indicates the times at which leptospirae are most likely to be demonstrated in the bloodstream, cerebrospinal fluid and urine.

Because leptospirae are extremely sensitive to an acid environment it is essential to test the pH of urine specimens immediately after collection; if necessary the pH should be adjusted to a level not less than 7.5. With any of these specimens three laboratory procedures can be used.

Microscopy

Films stained by silver impregnation techniques or wet films of the fresh specimen by dark-ground illumination should be examined, but are rarely positive except when the material has been harvested from sick animals infected by inoculation of specimens from the infected human.

Cultivation

Various media, all of which are heavily enriched, are available for attempted isolation of *L. interrogans* serotypes; several media should be inoculated with aliquots of the specimen, e.g. blood. Cultures are incubated at 30°C and examined every 3–4 days by dark ground microscopy; no culture attempt should be rejected as negative before 6 weeks incubation.

Animal inoculation

Intraperitoneal inoculation of young guinea-pigs or weanling golden hamsters with the relevant specimen is a sensitive diagnostic method; 3 days after inoculation abdominal paracentesis is performed and the aspirated fluid examined by dark-ground microscopy. If actively motile leptospirae are seen then cardiac puncture should be performed and the heart-blood used to inoculate fluid media. If at this stage the peritoneal aspirate shows no leptospirae then paracentesis is repeated daily for 4 more days.

With strains of leptospirae which are fully virulent to the experimental animal, the latter becomes febrile 4–5 days after injection and shows roughening of its coat and loses its

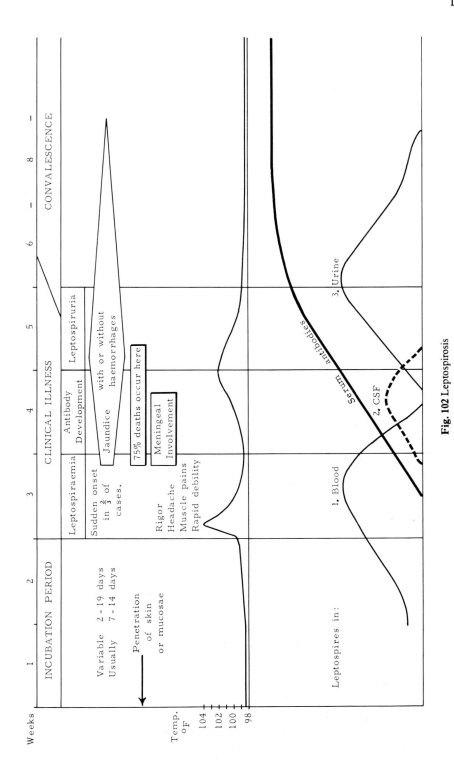

Fig. 102 Leptospirosis

appetite; jaundice appears by the 7th to 8th day and the animal dies within another 12–48 h.

Serological tests

Several techniques are available for the demonstration of specific antibodies in patients' sera. Agglutination tests with killed suspensions of the various serotypes are best suited for routine diagnostic purposes. It will be noted from Figure 102 that antibodies are rarely detected until the second week of the clinical illness, but a test should be carried out as early as possible so that a base-line titre for the patient can be determined and is available for comparison with later tests. As in all serological tests the most convincing evidence of recent infection is the demonstration of a rising titre when a second serum sample is tested 4–7 days after the first; however, a titre of 1 in 300 or more on first testing in a person with no occupational risk of previous subclinical infection can be regarded as diagnostic.

In people known to follow occupations which carry a risk of inapparent infection, titres of up to 1 in 100 can exist without clinical illness; in such cases repeated testing at 3–4 day intervals will show a steeply rising titre if they are infected whereas the titre will not rise in the absence of active leptospiral infection.

Miscellaneous infections

37

For those infections not dealt with in the processing charts and accompanying text the following procedures are adopted

WHOOPING COUGH

Nasopharyngeal or per-nasal swabs are plated out on Bordet-Gengou medium; in addition a 'cough plate' can be used — a plate of Bordet-Gengou medium is held 4–6 in in front of the patient while he is coughing and is thus inoculated with the expelled droplets. Bacteriological confirmation by isolation of *Bord. pertussis* is of a high order in the first 2–4 weeks of illness, especially in the catarrhal stage preceding the onset of whooping, but thereafter the isolation rate diminishes rapidly. Most laboratories incorporate a small amount of penicillin in Bordet-Gengou plates (6 units/12 ml of medium) thus reducing the growth of many other organisms and allowing any *Bord. pertussis* to be more easily recognised.

When whooping cough occurs in modified form in a child who has been actively immunised the prospect of isolating *Bord. pertussis* is even less likely regardless of the stage of the disease at which isolation is attempted.

Complement-fixation and agglutination tests may be applied to patients' sera after the third week of illness but such tests are rarely carried out; their role is the retrospective diagnosis of the atypical or missed cases at a stage of disease when isolation of the causative organism is unlikely.

CONJUNCTIVITIS

Swabs are processed as for the isolation of aerobic organisms from pus (Ch. 33) but additionally a heated blood agar plate should be incubated in an atmosphere with a 10% concentration of CO_2 to facilitate the isolation of *N. gonorrhoeae* and *H. influenzae*.

BRUCELLOSIS

During the febrile phases, blood culture should be performed; several such specimens should be obtained and the clinician should obtain Castaneda blood culture bottles. These incorporate a layer of liver-infusion agar along one of the narrow sides so that by tilting the bottle every 48 h the blood-broth mixture flows gently over the agar and the bottle is then reincubated in the upright position; the agar surface is inspected daily for developing colonies. Incubation should be continued for at least *4 weeks* before cultures are discarded as negative. During the chronic stage of the disease, blood culture is valueless; biopsy material, e.g. lymph glands, may be cultured successfully. The diagnosis during the afebrile phase is essentially by agglutination tests with the patient's serum against standardised suspensions of brucellae. Titres of 1 in 80 or more indicate present or past infection and a rising titre on testing a second specimen is conclusive of active disease.

Serological diagnosis is not always straightforward particularly in those exposed to infection by virtue of their occupation, e.g. veterinary surgeons and abattoir workers, and additional tests may be employed. Such tests, e.g. complement-fixation methods and radioimmunoassay tests, often assist in determining whether the infection is still active or whether the person has acquired immunity or has been treated successfully, but there is still no one test which will reliably indicate active disease. It is probable that this diagnostic dilemma will disappear in the next few years if the early success in eradicating the disease in Britain comes to full fruition.

GONORRHOEA

The feeble viability of the causal organisms outside the host's tissues demands that material for attempted isolation should be plated on to suitable media *immediately* and most S.T.D. clinics now have satellite laboratories on the premises and staffed with experienced technicians.

Smears are prepared directly from the discharge, and after staining by Gram's method these are examined for the presence of *intracellular* Gram-negative diplococci; it is essential also to culture discharges.

In the male, the meatus is cleaned with sterile gauze soaked in saline and the discharge collection with a sterile bacteriological loop and plated immediately; in the female, similar procedures are undertaken and in addition secretions from the cervix uteri must be examined and a vaginal speculum is essential.

Similar examination of rectal specimens should be undertaken in both sexes.

Selective media, e.g. Thayer Martin or Modified New York City media should be used; these contain various antibiotics which allow the growth of *N. gonorrhoeae* but discourage the growth of contaminating organisms including yeasts. Inoculated plates are incubated in a 10% CO_2 atmosphere for 48 h.

If direct plating is impracticable material should be inoculated into transport media, e.g. Stuart's or Amies's and dispatched immediately to the laboratory.

Serological diagnosis of chronic gonococcal infection relied for many years on the gonococcal complement-fixation test (GCFT) but false positive and false negative reactions have detracted from its original popularity.

Energetic research into serological techniques using highly purified fimbrial antigens or outer membrane proteins to detect antibodies is ongoing in the hope that specific and sensitive test systems will evolve.

Finally it is worth reiterating that all isolates of *N. gonorrhoeae* must be tested for β-lactamase production because of the increasing incidence of such strains throughout the world.

SYPHILIS

In the primary stage, serous exudate should be collected from the chancre after thorough cleansing of the area to reduce the population of other organisms; the exudate is collected in capillary tubes and delivered immediately to the laboratory for examination under the dark ground microscope. Refrigeration of such specimens is to be avoided since it immobilises *Tr. pallidum* and makes microscopic recognition very difficult. Similar procedures can be taken with exudate from skin eruptions and mucous patches in the secondary stage. At any time after 2–3 weeks from the onset of infection, diagnosis is best undertaken by testing the patient's serum for antibodies.

Early serological techniques relied on the detection of antibodies to cardiolipin and were either complement-fixation methods (the Wassermann reaction) or flocculation tests (e.g. the Kahn test); one such anti-cardiolipin antibody test still in use is the Venereal Diseases Research Laboratory (VDRL) slide test, but in common with such methods biological false positive reactions are encountered in patients, e.g. in atypical pneumonia and autoimmune diseases.

Several methods are now available for the detection of specific treponemal antibodies but only one, the *Treponema pallidum* Haemagglutination (TPHA) test, is widely used outside reference laboratories.

The TPHA test is easily performed, and syphilitic antibodies in a patient's serum are detected by their ability to agglutinate sheep red blood cells which have been coated with an extract of *T. pallidum;* it must be noted that the sera of patients suffering from other treponemal infections, e.g. yaws and pinta, also react positively.

The interpretation of serological tests requires close collaboration between venereologist and bacteriologist. Excellent guidance is found in several specialist texts, e.g. Robertson, McMillan & Young *Clinical Practice in Sexually transmissible Diseases* (Pitman Medical).

The possibility of both infections being acquired simultaneously must be borne in mind and since the incubation period of gonorrhoea is (3–9 days) in comparison with that of syphilis (3–6 weeks) then antibiotic treatment aimed at curing the gonococcal infection may mask the syphilitic infection which might proceed to the secondary stage without the development of a primary lesion.

Therefore all cases of gonorrhoea should be submitted to serological surveillance to ensure that they have not also acquired syphilis.

Bacterial sensitivity tests

38

There are few infective diseases or syndromes caused by a single bacterial species which is always sensitive to a particular antimicrobial agent; indeed with the exception of *Strept. pyogenes* which is constantly sensitive to penicillin, no species can be assumed to be sensitive to a particular antibiotic. In the previous edition of this book, *Pneumococcus* was included as a second example, but strains of pneumococci resistant to penicillin have been isolated in the last few years.

The need for laboratory control of antimicrobial therapy is widely recognised but perhaps in no other area of medicine does the practice fall so short of the preaching; in many areas clinicians, in hospital or in general practice, who seek laboratory confirmation of their diagnosis — let alone guidance in therapy — are in the minority and one is left with the opinion that many clinicians either never see patients suffering from infections or that they 'treat' such patients by guesswork. Such crystal-ball therapy is an abuse of antimicrobial drugs, nationally uneconomic, and not guaranteed to succeed.

Some of this abuse of antibiotics can be blamed on the laboratory which delays unnecessarily in reporting useful information; provided the clinician is aware of the limitations of rapid laboratory techniques he should appreciate that he can obtain

preliminary guidance of therapy at the same time as he receives a statement of the organisms isolated from the specimen — usually within 18–24 h of the material reaching the laboratory.

Before specifying available techniques several basic factors should be mentioned which may influence the results and which must be standardised.

Incubation time. If this is prolonged beyond the minimum required for growth of the organism under test, then there may be loss of activity of the antimicrobial agents which will then allow the growth of organisms which have suffered only bacteriostasis so that the antibiotic may appear to be less efficient than it actually is.

Medium. Provided that the medium is properly constituted the results of sensitivity tests are remarkably consistent and replicable; care has to be taken however since, for example, media used for sensitivity testing of sulphonamide activity must be free of substances inhibitory to that group of agents. Likewise, the addition of blood to the test medium will reduce the activity of antibiotics which are readily bound by protein.

pH of the medium. This too must be standardised to ensure reproducible results since, for example, the activity of the aminoglycosides is greatly enhanced by an

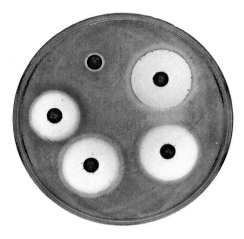

Fig. 103 These plates are identical except in the density of the inoculum of the strain under test; on the left the dense but non-confluent growth gives a truer picture of the activity of the antibiotic agents in the discs. The upper disc on the right has no activity against the organism under test and the reduced zone sizes to the other four antimicrobial agents emphasise the importance of inoculum size in disc diffusion tests

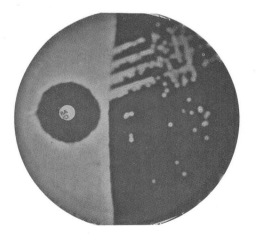
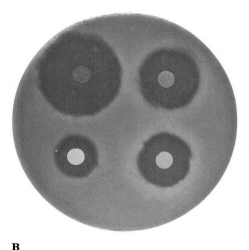

A   B

Fig. 104 A. This illustration shows the use to which an antibiotic may be put for diagnostic purposes; this crystal-violet blood agar plate was inoculated from a throat swab. A bacitracin disc was placed in the well inoculum and after 18 h incubation at 37°C a pure growth of β-haemolytic streptococci was obtained; sensitivity to bacitracin indicates that the strain belongs to Lancefield group A or more rarely to groups C and G, and thus facilitates rapid recognition of the groups most commonly pathogenic to man.
B. This illustration shows a subculture sensitivity test of the group A strain, and penicillin (pink), chloramphenicol (green), chlortetracycline (yellow) and streptomycin (white) are all active against the organism. Sensitivity tests are not usually performed against *Strept. pyogenes* since all strains are eminently sensitive to, and readily eradicated by, penicillin

161

alkaline medium whereas chlortetracycline is more stable and hence more active in an acid environment.

Inoculum size. An excessive inoculum will often result in smaller zones of inhibition around discs of antibiotics in the diffusion method and the inoculum should be such that it will produce dense but non-confluent growth; it is equally important that the inoculum should be seeded uniformly over the surface of the test medium and this is best achieved by flooding the surface with a suspension of the organism under test.

Disc diffusion technique

Diffusion techniques are less precise than tube dilution methods but are more speedily performed and allow the simultaneous testing of several drugs; furthermore, diffusion techniques can be employed in direct testing of the pathological material so that some indication of sensitivity can be given

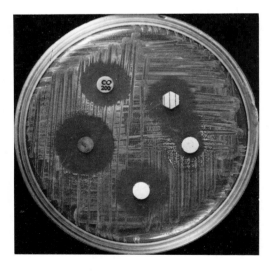

Fig. 105 Subculture sensitivity test. Nutrient agar plate seeded with *Sh. sonnei* and, after the application of antibiotic discs, incubated overnight at 37°C. The strain is sensitive to colomycin (blue), chloramphenicol (green) and streptomycin (white); it is also sensitive to paromomycin (orange) and chlortetracycline (yellow) but these agents are also *synergistic* as indicated by the obliteration of growth in the area between their respective zones

simultaneously with identification of the causative **organism** (Figs. 103–105).

Discs 6.25 mm in diameter are punched from Whatman No. 1 filter paper and in batches of 100 in screw-capped bottles are sterilised in a hot air oven at 150°C for 1 h. The desired antibiotic solutions are prepared in sterile distilled water and 1 ml added to each bottle; since 100 such discs absorb this unit volume each disc will contain approximately 0.01 ml. There is no general agreement as to the amount of a particular antibiotic contained in such discs but the following amounts per disc are useful in routine tests; penicillin 1 unit (10 units), chloramphenicol 10 μg, (30 μg), gentamicin 10 μg, tetracycline 10 μg, (30 μg), erythromycin 10 μg, trimethoprim 1.25 μg; where two disc contents are noted that in brackets is used to test the sensitivity of bacteria isolated from urine.

Commercially produced discs are frequently used but whether these or home made discs are employed, control cultures of stock organisms should be tested in parallel with clinical isolates to allow comparison of results.

In all diffusion tests of activity the depth of medium must be standardised by using a constant volume in a Petri dish of constant diameter and with a flat surface; such plates must be poured on a horizontal surface. The inoculum may be spread uniformly over the surface with a swab from a broth culture or the suspension may be flooded over the whole surface with a Pasteur pipette; thereafter, the plate is tilted and excess inoculum removed with the pipette. The inverted plates are left to dry on the bench for 1 h before discs are placed on the surface with sterile forceps. Ideally a period of 3 h prediffusion should be allowed before the plates are incubated. Diffusion tests may be performed using reservoirs other than discs for the antibiotics, e.g. a ditch may be cut from the medium and replaced with the antimicrobial agent mixed with agar, or cylinders (porcelain or steel) may be placed on the medium and the required

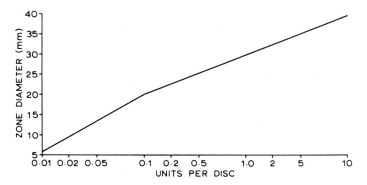

Fig. 106 Graph showing zone diameters of inhibition for concentrations of penicillin

agent pipetted into these.

Standard graphs have been prepared for each antibiotic which show the zone diameters of inhibition given by various concentrations of the antibiotic against a standard organism and these graphs allow the results of disc diffusion tests to be interpreted quantitatively.

The graph for penicillin is shown in Figure 106, and by measuring the zone of inhibition of an organism under test with a disc of known activity one can calculate the

sensitivity of the organism to penicillin in comparison with that of the standard organism.

Tube dilution technique

Tube dilution testing requires the preparation of two-fold dilutions of the antibiotic in a suitable fluid medium; a constant volume of the organism is added to each tube and a control tube containing no antibiotic is included in each test. After incubation at 37°C

Fig. 107 All tubes contain an identical volume of broth (the slight differences in level are caused by variation in the internal diameters of the tubes). Doubling dilutions of an antibiotic were prepared in the series so that tube 1 contains the highest dilution and tube 10 the lowest dilution of antibiotic; tube 11 is a control tube of broth without antibiotic.

To each tube was added an identical volume of an overnight broth culture of *Shigella sonnei;* after 18 h incubation at 37°C, macroscopic examination revealed that growth had occurred in tubes 1–5 and 11.

The bacteriostatic concentration of the antibiotic thus lies between the amounts present in tubes 5 and 6; to determine the *bactericidal* concentration, aliquot volumes from each of tubes 6–10 were seeded on to a nutrient agar plate. After 18 h incubation, growth was noted on the sector seeded from tube 6 and scanty growth on that inoculated from tube 7.

The *bactericidal* concentration of the antibiotic lies between those concentrations present in tubes 7 and 8

for 18–24 h the tubes are examined for turbidity and that with the highest dilution showing no visible turbidity is the *bacteriostatic* concentration (Fig. 107). The *bactericidal* concentration can be determined only by subculturing from those tubes without visible growth and transfer may be to agar plates or into broth devoid of antibiotic; the highest dilution yielding no growth on subculture is the bactericidal concentration (Fig. 108). The tube test is too expensive of time, labour and materials for general purposes and it is normally reserved for testing the sensitivity of slow-growing organisms, particularly *M. tuberculosis*.

Fig. 108 Determination of bactericidal concentration. This nutrient agar plate was seeded from tubes 6–10 in the dilution series and after 18 h incubation, growth appeared only on these sectors inoculated from tubes 6 and 7

Assay of antibiotic levels in body fluids

There are three situations when it is necessary to establish the concentration of an antibiotic in human body fluids; firstly, *to confirm that adequate levels of antibiotic are being attained,* e.g. in the CSF of a patient with meningitis or more generally when an antibiotic is being

administered orally to a patient whose intestinal tract may be deficient in its absorptive ability; secondly, *to ensure that blood levels do not exceed the limits* at which certain antibiotics, e.g. gentamicin, may cause nerve deafness; and finally, in *the investigation of a new antimicrobial agent,* assay of levels in body fluids allows the determination of its adsorption, distribution within the body and its rate and route of excretion.

Assay may be performed by either a diffusion technique or tube dilution method; if the material to be assayed for antibiotic content is likely to be contaminated with microorganisms these must be removed by filtration or some other method before tests are made by the tube dilution method. In this method doubling dilutions of the material are prepared in nutrient broth and each tube is then seeded with equal volumes of a standard organism of known sensitivity to the agent being assayed. A control set of tubes is set up containing not only the antimicrobial agent under assay but also a fresh sample of the same body fluid free from the antibiotic, so that any natural bacteriostatic activity of the particular body fluid occurs in the control as well as in the test series of tubes. Both sets of tubes are then incubated, and by comparing the dilutions which give inhibition of the growth of the standard organism one can estimate the amount of antimicrobial agent in the patient's fluid.

Assay by the diffusion method is carried out using porcelain cylinders in place of discs; once the porcelain cylinders have been sealed into plates seeded with a test organism of known sensitivity to the antibiotic being assayed they are filled with dilutions of the fluid under test.

After incubation the significance of the zone diameters of inhibition are read from a standard graph prepared for the particular antibiotic.

One advantage of the cylinder diffusion method is that contaminated material, even faecal suspensions, can be assayed without attempting to get rid of microorganisms.

These brief comments silhouette the microbiological methods of assaying the levels of antibiotics in body fluids. Many other methods, chemical as well as biological, are available and for details of these a suitable reference book should be consulted; the author frequently consults the most recent edition of Garrod, Lambert & O'Grady, *Antibiotic and Chemotherapy,* published by Churchill Livingstone.

Prophylactic immunisation

39

Immunity is the ability of the animal body to resist infection with microorganisms and the harmful effects of their toxins; reference has been made to immunisation procedures in various chapters in the second section of the text and in this chapter further synoptic notes on immunisation are offered.

INNATE IMMUNITY

Every animal, including man, possesses natural barriers that protect it against microbial attack; the skin and mucous membranes offer physical and chemical defences. For example, few bacteria can breach unbroken healthy skin — the exceptions are *T. pallidum* and *L. interrogans*; similarly healthy mucosal surfaces are well-equipped to trap attempted invaders in mucous secretions. Chemical defence of the skin is provided by the production of long-chain fatty acids, e.g. oleic acid, which maintain an acid pH of around 6 on almost all of the body skin. Lysozyme, a naturally produced enzyme in almost all tissues, provides an effective defence, e.g. in tear fluid bathing the conjunctivae.

Organisms which escape these physical and chemical gauntlets immediately have to contend with phagocytes, either in the bloodstream or the fixed phagocytes of the reticulo-endothelial system.

Other natural but somewhat more specific defence mechanisms include the antagonism displayed by the commensal flora against potential invaders; e.g. *Lactobacillus acidophilus* (Döderlein's bacillus) lives commensally in the vagina and there creates a highly acid secretion by fermenting the glycogen of the vaginal epithelium; similarly lactobacilli which are commensal in the gut probably exert a protective effect since, when they are eliminated by the use of broad-spectrum antibiotics, severe and sometimes fatal infections occur. These are usually caused by strains of *Staph. pyogenes* which have become resistant to the broad-spectrum antibiotics, or by *Candida albicans*.

Properdin, a high molecular weight protein which is present in normal serum, is another component of natural immunity, but its activity is directed against Gram-negative bacteria; other basic proteins with an antibacterial action are produced from tissue and blood cells which have been damaged by infection, e.g. spermine, which is lethal to tubercle bacilli.

ACQUIRED IMMUNITY

In addition to the natural protection against

infection noted above, an individual may acquire immunity of a highly specific type, i.e. protection against a particular bacterium or its toxins.

Such immunity may be acquired naturally or artificially.

NATURALLY ACQUIRED IMMUNITY

This type of immunity may be obtained *actively*, i.e. when the individual's tissues produce antibodies specific for an organism which has infected them, and such active immunity is usually long-lasting.

Natural immunity may also be acquired *passively* by transplacental donation of antibodies from the mother to the foetus; such passive naturally-acquired immunity is of short duration and rarely lasts more than 3–6 months, since the antibodies are foreign to the infant's tissues and are eliminated.

ARTIFICIALLY ACQUIRED IMMUNITY

Our profession mimics naturally-acquired immunity, firstly and most frequently by administering antigens (by various routes) to stimulate the patient's tissues to produce specific antibodies. As in active immunity acquired naturally, the protection thus obtained is long-lasting in comparison with the second method of artificial immunisation, i.e. the injection of antibodies which have been obtained from some other host, where protection rarely lasts for more than 4–6 weeks.

Prophylactic immunisation, i.e. the acquisition of immunity to microbial infection by artificial methods, is one of the most spectacular successes in modern medicine and has been responsible for the virtual elimination of mortality in several infections.

The vast majority of the agents which we now use to induce immunity artificially have been introduced in the present century as a logical development of microbiological research, but active artificial immunisation had been practised, particularly against smallpox, for very many years. The earliest recorded account was of variolation which was introduced to Britain 250 years ago by Lady Montagu, who was the wife of the British Ambassador in Turkey; variolation, i.e. the administration by scarification of material from a smallpox vesicle into the skin of a healthy subject who then usually suffered and survived a modified form of the disease, had been common practice in Turkey for decades. However, the disease resulting from variolation was often severe and sometimes fatal and in addition variolated subjects frequently acted as sources of epidemic infection; thus variolation was made illegal in Britain in 1840. Perhaps the main reason for banning variolation was that Jennerian vaccination, introduced in 1798, was much more successful and certainly less dangerous.

The only other attempt at prophylactic immunisation which preceded the discovery of microorganisms was in the case of measles; here on a more restricted basis Francis Home reported in 1758 his method of implanting, by skin incision, threads soaked in material from measley children in an endeavour to produce a modified form of the illness which would result in protection from the natural form of the disease which at that time carried a high mortality rate.

Fortunately the introduction of the hypodermic syringe (1853) preceded the discovery of microorganisms and our ability to offer active and passive immunisation to microbial infection.

Safety factors in artificial immunisation

The syringe. Like most other discoveries the hypodermic syringe has its disadvantages, and these are associated with the possible transmission of infection. Unless a syringe and needle are *sterile* then sepsis is a constant risk following injection not only of immunising agents but of any fluid; an even more frightening risk is that of the transmission of the virus of serum hepatitis from person to person if a syringe and needle are used communally, e.g. at immunisation clinics. The lengthy incubation period, on average 80 days (range = 40–160 days), following the injection of the causal virus (Virus B) frequently prevents the correlation of the infection with a particular injection unless an outbreak occurs, e.g. when several patients develop serum hepatitis and all have a history of venepuncture at a particular clinic on a given day; serum hepatitis occurs when even minute amounts of blood, as little as 0.001 ml, are transmitted from a case or carrier to another individual, e.g. by a lancet used for finger pricking to obtain small volumes of blood for blood cell counts.

Perhaps it is not unnatural that the incidence of sepsis and serum hepatitis following on injection is not known since such iatrogenic episodes are rarely published.

However, both can be avoided if *a separate sterile syringe and needle* are used for each individual receiving an injection.

Human errors. The production of immunising agents is rigorously controlled and tests of the safety and potency of each batch of an agent are performed before the agent is issued for use. However, human frailty has led to occasional disaster; in the early days of BCG production when, in Germany, the vaccine was produced in a laboratory where virulent, human-type tubercle bacilli were also being grown, more than 200 babies were accidentally given virulent bacilli instead of BCG and almost one-third of the babies died of acute tuberculosis. Much more recently, and when more stringent procedures have been instituted for the control of vaccine, catastrophes still occur, e.g. the distribution of 'killed' poliomyelitis vaccine which contained live virulent polioviruses.

Complications of prophylactic immunisation

When an infection is endemic and rampant in a community and then immunisation against the infection becomes available, the risk of the complications of immunisation is acceptable to that community; however when the particular infection has been eradicated, primarily as a result of active artificial immunisation, a stage is reached when more people suffer side-effects and death as a result of preventive inoculations than the number who die from the actual disease. At this point, the risk of prophylactic immunisation becomes unacceptable so that fewer people may be protected artificially and the community then becomes more and more susceptible to the natural disease.

Only by a continuing programme of health education and stressing the need for immunisation can we hope to remain free from the return, in an endemic form, of many infectious diseases such as diphtheria and poliomyelitis.

As examples of the complications of active immunisation we can consider briefly smallpox vaccination, pertussis immunisation and provocation poliomyelitis.

Smallpox vaccination. Smallpox has now been eliminated globally and only a few countries persist in demanding certificates of vaccination for those entering their territories. Apart from localised sepsis at the site of vaccination the serious complications are post-vaccinial encephalitis and generalised vaccinia and both of these were minimal when primary vaccination was undertaken after the patient's first birthday and before the fifth birthday.

Tragically we still see occasional cases of severe complications following the unnecessary administration of smallpox vaccine in the UK!

Pertussis immunisation. The vaccine used comprises killed strains of *Bord. pertussis* possessing the antigens associated with the serotype prevailing in the community; the introduction of an efficient vaccine into the UK for mass immunisation in childhood was followed by a dramatic reduction in the incidence of the natural disease although the

3–4 yearly epidemicity was still in evidence.

The suspected association, in a few instances, between the administration of pertussis vaccine (as part of the Triple vaccine) and convulsions with subsequent evidence of brain damage prompted an aversion against pertussis vaccine in the mid-1970s; the fall in acceptance rates was followed, inevitably, by increases in the incidence of infection to levels not seen since 1955.

Prospective investigations in two British cities reveal so far that the incidence of 'brain damage' is infinitesimal compared with that following the natural disease *provided* that children with a prior history of convulsions or those with a first-degree relative with a history of epilepsy are not immunised. Similarly a child should not be immunised if at the appointed time it is suffering any respiratory infection; certainly if an adverse reaction has followed a previous dose of vaccine no further doses should be given.

Insufficient emphasis is given to the pulmonary complications of whooping cough, which are often not evident until later in life; these can be eliminated if the child is protected with pertussis vaccine. It should be noted that a double toxoid preparation is available to give protection against diphtheria and tetanus and should be sponsored for use when parents do not wish their child to receive the pertussis component of the Triple Vaccine.

Provocation poliomyelitis. The possibility of poliomyelitis being provoked by immunisation procedures is now a matter of history and will remain so as long as we ensure that children receive poliomyelitis vaccine early in their immunisation programme.

Efficacy of prophylactic immunisation

With certain agents the high degree of protection is easily seen but in general the results of laboratory tests in experimental animals must be confirmed by controlled field trials in the human population; in the past this ultimate proof has not always been undertaken and we have occasionally been misled by animal experiments.

As an example of the importance of field trials in evaluating the efficacy of vaccines we can note in summary fashion the history of immunisation against the enteric fevers.

Typhoid vaccines. Immunisation against typhoid fever was introduced in 1897 and in 1916 the vaccine also incorporated antigens to offer protection against infection with *S. paratyphi A* and *S. paratyphi B*. This typhoid vaccine was heat-killed and phenol was used as the preservative — 'phenolised vaccine'; it had never been subjected to properly controlled evaluation in man.

In 1934 a new type of typhoid vaccine — 'alcoholised vaccine' — was reported to have a much higher protective value for mice by virtue of its Vi or virulence antigen which stimulated the formation of Vi antibody; in 1941 it was reported that human beings who had received alcoholised vaccine frequently possessed Vi antibodies which were rarely produced in men receiving phenolised vaccine.

However, the relative efficacy of these two types of typhoid vaccine was not determined until field trials were undertaken in Yugoslavia in 1954; as a result it was shown that the incidence of typhoid fever in a population receiving phenolised typhoid vaccine was 6.1 per 10 000 as compared with an incidence of 14.1 per 10 000 individuals who had received the alcoholised vaccine. The incidence of typhoid fever in a control group who received a 'vaccine' comprising *Shigella flexneri* was 19.2 per 10 000 individuals.

Thus it became obvious that no matter how efficient the alcoholised vaccine is in protecting mice against infection it had a very low protective value for men.

Further field trials in Guyana involved the use of phenolised vaccine and a new acetone killed vaccine. The latter has an even greater protective effect against typhoid fever since the attack rate in the group receiving acetone-killed vaccine was only 1 per 10 000.

In neither of these trials was the efficacy of the paratyphoid components assessed and the official recommendation, for those requiring protection against typhoid fever, is that a monovalent acetone-killed typhoid vaccine should be given; this can be given intracutaneously and local side effects are significantly less than with earlier vaccines.

BCG vaccine. Whilst it is accepted that several factors are involved in the elimination of tuberculous infection, e.g. early detection and treatment of cases, tracing of contacts, provision of dairy herds free from bovine infection and/or pasteurisation of milk supplies, etc, the role of preventive immunisation with BCG or Vole vaccines is important.

The evaluation of these vaccines in field trials in Britain has been perhaps the most exciting work in post-war endeavours in prophylactic immunisation. These trials, instituted by the Medical Research Council in 1949, have clearly shown that either vaccine gives an 80% reduction in the incidence of natural disease and in particular they protect against the more severe forms of infection, i.e. tuberculous meningitis and miliary tuberculosis, which in the pre-antibiotic era were almost invariably fatal.

Diphtheria toxoid. Those old enough to remember the endemic and potentially lethal nature of diphtheria before mass immunisation was undertaken in 1941 recall, with relief, the dramatic reduction in

incidence and the even more spectacular fall in mortality rates from diphtheria (Fig. 109). No field trials were needed to display the efficacy of diphtheria toxoid and diphtheria, at present, is part of our folk lore. However, the reduced acceptance rates of vaccines associated with the pertussis scare in the mid-1970s means that, as a nation, we are breeding a young population unprotected against diphtheria and since importation is always a prospect we are increasingly at risk from epidemics; such a tragedy could readily be compounded since few medical graduates under the age of 50 have seen cases of diphtheria and not only is a delay in diagnosis guaranteed to increase the risk of death but also allows the undetected case to remain out of isolation with the prospect of rapid spread of the disease to other susceptible individuals.

Tetanus toxoid. Similar evidence for the high protective value of tetanus toxoid has been gathered over the last 50 years during wars; the low incidence of tetanus in civilian life makes it impracticable to undertake controlled field trials in most countries. Nevertheless, when one compares the incidence of tetanus in battle casualties in three periods 1) before any form of immunisation was available, 2) when passive immunisation could be given after wounding, and 3) in the Second World War when active immunisation was performed, then the spectacular drop in incidence can be noted (Table 12).

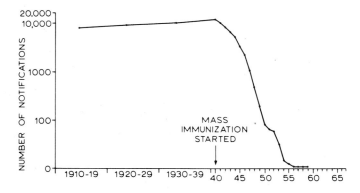

Fig. 109 Diphtheria notifications: Scotland 1910–1967

Table 12. Incidence of tetanus in battle casualties

	Cases of tetanus/1000 casualties
Pre-immunisation	3–10*
Passive immunisation (after wounding)	1.5
Active immunisation (before wounding)	0.1

*Variation in incidence depended on the type of terrain on which fighting took place, i.e. highest on cultivated manured land.

Tetanus should not be associated solely with deep wounds containing necrotic tissue, since in civilian life the injury allowing implantation of the causal spore is usually trivial, e.g. a rose-thorn prick in a finger.

Pneumococcal vaccine. This is the most recently introduced bacterial vaccine and comprises a saline solution of highly purified capsular polysaccharides extracted from the 14 most prevalent serotypes of pneumococci. Its value has still to be fully assessed; it is administered as a single dose given subcutaneously.

Its use would appear to be beneficial to patients who have undergone splenectomy since such individuals are known to have a high risk of suffering severe pneumococcal infections; other categories who may benefit are those with sickle-cell disease, chronic liver disease or primary immunodeficiency. The response to this polyvalent pneumococcal vaccine is poor in children under 2 years of age.

Passive artificial immunisation

The brief duration of passive artificial immunisation has already been noted and the short-term nature of such protection is dictated by the fact that the antibodies have almost invariably been derived from animals other than man and the human tissues recognise the complete foreignness of the animal globulins which are rapidly eliminated in a few weeks.

An additional disadvantage of passive immunisation with antisera derived from animals is the risk of hypersensitivity reaction to animal serum, e.g. immediate anaphylactic shock may occur in individuals who frequently have a personal or family history of allergy, or serum sickness may occur as a delayed phenomenon some 1–2 weeks after administration of the antiserum. Such hypersensitivity reactions are now less common since tetanus antitoxin is nowadays derived from human volunteers who have been hyperimmunised with toxoid and then donate units of blood for the production of antitoxin; this reduced risk is however no argument against widespread *active* immunisation of populations with tetanus toxoid.

Immunisation schedules

The diseases against which active immunisation is presently available are listed below; it is indeed fortunate that for ecological reasons no one country requires to subject its citizens to protection against all the diseases listed. Diseases against which people living in Great Britain should be protected are indicated by asterisks.

Cholera	Rubella*
Diphtheria*	Tetanus*
Influenza	Tuberculosis*
Measles*	Typhoid fever
Pertussis*	Typhus fever
Plague	Yellow fever
Poliomyelitis*	

The triple vaccine has priority in Britain because although pertussis is not frequently encountered in the first year of life more than half of the deaths from whooping cough occur before the first birthday.

We must reiterate the need to pursue constantly intelligent methods of health education, so that mothers appreciate that

171

only by continued artificial immunisation can we hope to prevent many communicable diseases from re-establishing themselves in Britain; this is particularly pertinent in the case of infections, such as diphtheria, which were once endemic but are now so rare that most young parents are aware of them only from family folk-lore.

Many authorities recommend that rubella vaccine need only be given to girls but if our endeavour is to eliminate rubella from our community then boys should also be immunised with that vaccine; a national campaign has been mounted in the USA to eliminate measles and the incidence of the disease has fallen dramatically in the last few years. One of the obvious problems with such energetic national endeavours is that, unless protection following immunisation is lifelong, then we may be simply postponing the incidence of essentially childhood infections, such as measles, into adult life and we know that, in general, such infections are more severe when suffered in adulthood.

Active immunisation can also be offered against certain other diseases but such protection is restricted to individuals at particular risk, e.g. those with an occupational risk of acquiring infection; these infections are listed below.

Anthrax	Mumps
Brucellosis	Q-fever
Leptospirosis	Tularaemia

In constructing a policy for immunisation in a particular country, priority must be given to preventing those infections which feature most frequently as causes of morbidity and mortality; for example, in Nigeria the incidence of tuberculosis dictates that immunisation against that infection should be undertaken in the first four months of life.

In many developing countries, and especially those with an essentially rural and/or nomadic population, there are tremendous logistic problems in ensuring that children receive immunising agents; even in more sophisticated communities the tedium of repeated visits to immunisation clinics has contributed to a fall-off in the proportion of the population who seek active immunisation and has stimulated the search for combined prophylactic agents as well as the creation of immunisation schedules which offer maximum protection with the minimum of inconvenience to mother and child.

The schedule outlined in Table 13 is suggested for countries where public health services are satisfactory.

Table 13. Immunisation schedule

Age	Visit	Vaccine	Time interval
3–10 months	1	Triple* & oral polio	
	2	Triple & oral polio	6 weeks
	3	Triple & oral polio	4–6 months
2–4 years	4	Measles	
5 years	5	Dip./Tet. & oral polio	
9 years	6	Dip./Tet.	
11–13 years	7	Rubella vaccine	
15 years	8	BCG†	

*Triple vaccine comprises pertussis, diphtheria and tetanus antigens
†For tuberculin-negative children

SECTION 4

Protozoology

Fortunately protozoal infections are less common in Britain than in certain other, mainly tropical countries; this section offers some information about a few of the more important species of protozoa that are pathogenic to man.

Protozoa have several features in common with bacteria but differ from the latter so significantly in several respects that it is suggested that protozoa should be regarded as closer to mammals than to bacteria.

Like bacteria they are unicellular microorganisms but are usually *much larger* (2–100 μm). Whilst some protozoa reproduce by simple binary fission, others undergo *schizogony* or multiple fission; i.e. as the cell grows the nucleus divides repeatedly so that the cell becomes multinucleate, and only when growth ceases does cytoplasmic fission occur with the emergence of numerous uninucleate *merozoites*. *Endodyogeny*, i.e. the formation of two similar daughter cells within the parent cell, is characteristic of certain protozoa, e.g. *Toxoplasma* species. Sexual reproduction occurs in some parasitic protozoa, e.g. malaria parasites, where zygotes are formed from the union of non-motile macrogametes with small, motile microgametes.

Protozoa also differ from bacteria in usually ingesting foodstuffs in particulate form, and digestion proceeds within food vacuoles whereas much of the enzymatic activity of bacteria occurs at the cell surface; the organisation of the protozoal cell is mammalian in nature since within the cytoplasm there is a nucleus which contains DNA combined with protein to form chromosomes. The mitochondria of the protozoal cell resemble those of higher animals in structure and function, and another similarity with mammalian cells is that protozoa can resist the action of antibiotics at concentrations which destroy bacteria.

Many protozoa have complex life cycles both in the human or animal host and in insect vectors; several species can form cysts, a stage in the life cycle when the organism is surrounded by a definite membrane. Cysts represent a resting phase in which the organism is much more resistant to adverse environments than are trophozoites. Mitosis may occur in some protozoal cysts, e.g. those of *Giardia lamblia,* so that two or more trophozoites are produced during excystation.

Entamoeba histolytica

40

Several species of amoebae can inhabit the human gut, but only *Entamoeba histolytica* is pathogenic to man; it can also infect other primates. Man is infected by ingestion of the cyst form, either by direct contact with a carrier or case or indirectly from contaminated water supplies or from food contaminated by an infected food-handler. Insects can also transmit cysts from excreta to foodstuffs.

After ingestion by the host, those cysts which escape the gastric juices pass into the small intestine where they excyst or hatch to form vegetative cells, i.e. trophozoites, which colonise the large intestine, particularly the caecum and pelvic colon (Fig. 110).

Trophozoites may remain free-living in the intestinal lumen and in turn encyst and are excreted in the faeces, often in large numbers; alternatively they may invade the intestinal wall and cause ulceration. Classically, amoebic dysentery runs a chronic course in comparison with bacillary dysentery, and from the primary site in the intestine *Ent. histolytica* may spread by the lymphatics and bloodstream to the liver to give rise to amoebic abscesses; more rarely metastatic spread to other organs, e.g. lungs and brain, may occur. *Cysts are never found in infected tissues.*

The *trophozoite* measures 20–40 μm in diameter and contains a single nucleus; this vegetative form is highly active, divides by binary fission and ingests red and white blood cells and tissue cells (Figs. 111 and 112).

The *cyst form* of *Ent. histolytica* (Fig. 113) is less than half the diameter of the trophozoite and is spherical, with a single smooth wall, and translucent; cysts contain one, two or four, but *never more than four,* nuclei. Thick ovoid or rod-shaped structures known as chromatoid bodies or chromidial bars may also be seen.

LABORATORY DIAGNOSIS

Although cultural and serological techniques are available for the diagnosis of amoebiasis, this can almost invariably be made by microscopic examination of specimens of faeces for trophozoites or cysts, either in wet films or stained preparations; similarly, a search can be made for trophozoites in pus from abscesses or material obtained by biopsy.

Faeces must if possible be examined within an hour of the specimen being passed, but if this is impossible then the specimen should be fixed in four or five times its own volume of polyvinyl alcohol or placed in a refrigerator;

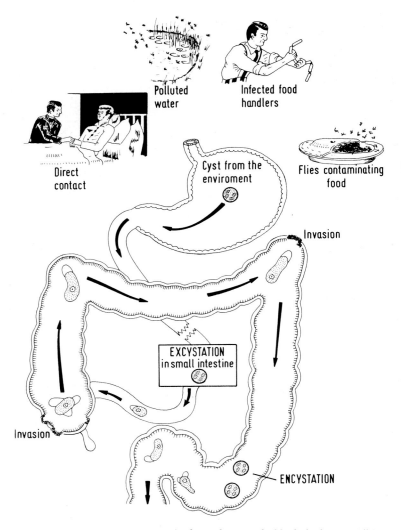

Polluted water

Infected food handlers

Direct contact

Cyst from the enviroment

Flies contaminating food

Invasion

EXCYSTATION in small intestine

Invasion

ENCYSTATION

Fig. 110 Amoebiasis. The trophozoite form of *Entamoeba histolytica* is responsible for infection within the human host and the cyst form, which is very much more resistant to adverse environmental conditions, transmits infection from man to man by direct contact with a carrier or case of amoebic dysentery. Indirect spread via polluted water supplies or food infected from food handlers who are carriers or by flies contaminating food after feeding on faeces from a carrier or case also occurs

on no account should a 'delayed' specimen be kept warm since the amoebae will disintegrate rapidly. Naked eye examination of the specimen should not be ignored and any visible mucus should also be examined microscopically since amoebae usually adhere to it.

When preparing a sample for microscopy, the microscope slide and the saline used for mixing the specimen must be heated to body temperature, since amoebae become sluggish or immobile in a cold environment. A small portion of the specimen is mixed with warm saline on a warmed slide and *the preparation*

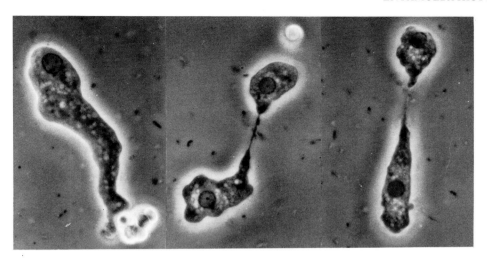

Fig. 111 *Entamoeba histolytica.* Time-lapse photography under phase contrast, showing a trophozoite dividing; the character of the nuclei should be noted, particularly the ring of fine chromatin granules lying on the nuclear membrane and the small karyosome which is situated centrally. These features, which are usually seen even more clearly in stained preparations, allow differentiation from the commensal *Entamoeba coli* where the karyosome is not centrally situated and the chromatin is coarser
× 750

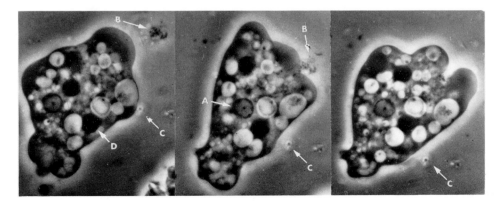

Fig. 112 *Entamoeba histolytica.* Time-lapse photography under phase contrast. The point at C is a marker against which the movement of the trophozoite can be judged; at point B are particles towards which the trophozoite progresses and eventually engulfs. D indicates a recently ingested red blood cell and at A the characteristic nucleus is evident.
× 750
The cells in both Figure 111 and Figure 112 were harvested from a culture in Jones' medium

177

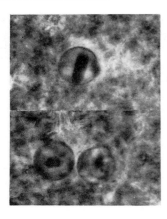

Fig. 113 *Entamoeba histolytica.* A phase contrast view of the cystic stage; these are spherical with a thin smooth wall containing a homogenous cytoplasm within which the chromatoid bodies show as dark objects.
× 750

should not be too thick or the amoebae may be obscured by debris; the preparation is best viewed by phase contrast microscopy, and the presence of amoebae containing red blood cells allows the clinical diagnosis to be clinched with confidence.

When a search is to be made for cysts it is customary to concentrate these from a sample of faeces by the zinc sulphate flotation technique, which allows the cysts to float whilst most of the faecal matter sediments (Fig. 114). Since the number of cysts fluctuates in stools, at least three specimens, collected on separate days, must be examined and found to be negative before an individual patient is considered free from infection

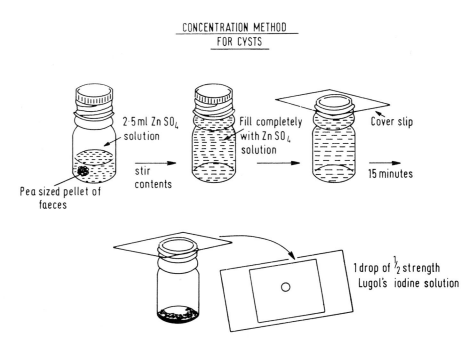

Fig. 114 Concentration of protozoal cysts in faeces: flotation method. A 33% solution of zinc sulphate ($ZnSO_4$) is used and 2·5 ml is placed in a 5 ml screwcap bottle; a sample of faeces is thoroughly emulsified in the solution and then the bottle is filled to the brim with $ZnSO_4$. Any large particles which float to the surface are removed and the solution replenished to be level with the top of the bottle; a coverslip is then laid in contact with the fluid surface.

After 15 min the coverslip is removed and placed, wet surface downwards, on a microscope slide on which a drop of ½ strength Lugol's iodine has been deposited

Serological diagnosis

Immunodiagnostic tests on patients' sera are best carried out in Reference Laboratories, but a significant percentage (35–45%) of patients suffering only *intestinal* amoebiasis will show no serological evidence of infection, which may mislead the clinician; on the other hand a patient suffering amoebic liver abscess almost always gives a positive serological result.

The treatment of amoebiasis, both intestinal and extra-intestinal, has been revolutionised by the use of metronidazole.

Giardia lamblia

41

This flagellate protozoon inhabits the duodenum and jejunum, and although it can be detected in the faeces of healthy individuals it is also incriminated as a cause of epidemic enteritis in young children, particularly those living in residential or day nurseries. Adults are certainly not immune and on occasion suffer severe diarrhoea with weight loss, often over some weeks, before the diagnosis of giardiasis is considered; water-borne outbreaks have been reported, which underlines the fact that the cysts of *Giardia lamblia* tolerate normal chlorination procedures. Metronidazole is now the therapeutic drug of choice.

G. lamblia reproduces by binary fission and exists both as a trophozoite and in a cyst form.

The *trophozoite* is flat and pear-shaped

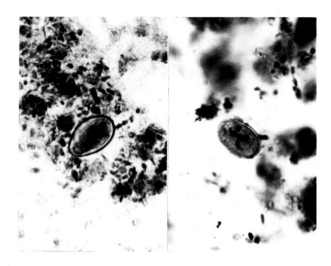

Fig. 115 *Giardia lamblia*. These two preparations were made from faecal specimens obtained from two children who had recently suffered from giardiasis. Both show a single cyst form of the parasite each of which contains two organisms formed by sub-division. The parallel axostyles can be noted in the cyst on the right
$\times$ 1000

when viewed dorsally and in the lateral view is crescent-shaped, the concave ventral surface acting as a sucker to fasten the organism to the intestinal epithelium; it measures approximately 10×15 μm. Two nuclei are present near the broader anterior end and a comma-shaped body, the median body, lies posterior to the nuclei. Four pairs of flagella originate on the ventral surface anteriorly, and the trophozoite has a characteristic rolling movement when seen in wet preparations.

Cyst forms are oval and somewhat smaller than the trophozoite, and possess two or four nuclei; parallel axostyles may also be noted (Fig. 115).

Giardia species which are found in the intestine of other mammals are morphologically similar to *G. lamblia* but are given specific status according to the host which is parasitised, e.g. *G. canis* in dogs; it is not known whether transmission from animal to human host, or vice versa, occurs.

Laboratory diagnosis depends on the microscopic detection of trophozoites and/or cysts in faecal specimens; the zinc sulphate flotation method for cysts can also be used.

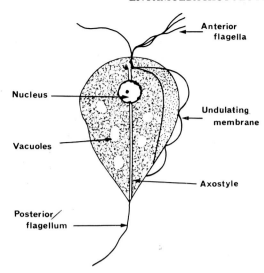

Fig. 116 Line drawing showing the essential features of *Trichomonas* species

TRICHOMONAS VAGINALIS

This is the only member of the genus which is pathogenic to mankind, although others are common parasites in the intestinal tract of many vertebrates and one other, *Trichomonas foetus,* infects cattle and causes early abortion.

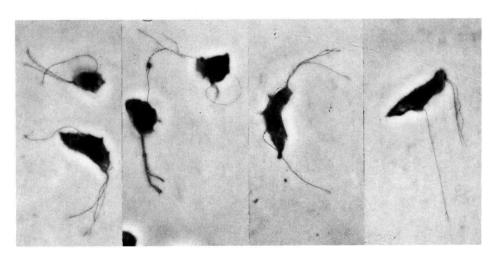

Fig. 117 *Trichomonas vaginalis.* Time lapse photography under phase contrast of *T. vaginalis* from a culture. The characteristically jerky movements of the organism and rotation on its long axis make it difficult to record certain characteristics in wet preparations although the undulating membrane can be noted in the lower organism in the first frame and also in that in the third frame. At least two and sometimes three of the four anterior flagella can be noted in all of the organisms
$\times$ 1000

T. vaginalis is ovoid, measuring 10×15 μm and possesses four anterior flagella and an undulating membrane which extends halfway down one side of the organism; the single nucleus is situated near the broader anterior end (Fig. 116). *T. vaginalis* does not form cysts, but the trophozoites can survive outside the host for a day or two provided that the environment is suitable, i.e. cool and moist.

A similar or perhaps identical trichomonas has been isolated from urethral and prostatic secretions in men and from the faeces of both sexes; there is, however, some doubt as to its pathogenicity in these situations but there is no doubt that *T. vaginalis* causes vaginitis which may occur as an entity or in association with gonococcal infection.

Laboratory diagnosis

This is usually undertaken by microscopic examination of wet films of secretions, (Fig. 117) but *T. vaginalis* can also be cultured in suitable media.

Toxoplasma gondii

42

Toxoplasma gondii, the causal organism of toxoplasmosis, obtained its specific name from the fact that one of the earliest observations of the infection was in the gondi, a North African rodent, in 1908; five years later the first properly documented human case was reported. *T. gondii* is now known to parasitise many species of mammals and birds, and is exceptional amongst the protozoa in being able to develop in all kinds of host cells with the exception of the non-nucleated red blood cell.

The trophozoites of *T. gondii* are crescent-shaped, 6×2 μm, with a single oval nucleus, and in the acute stage of infection the trophozoites multiply rapidly within the host's cells; such rapidly reproducing forms are referred to as tachyzoites, and reproduction is by endodyogeny, i.e. the formation of two daughter cells within the parent. Later, in the chronic stage of clinical infection, only 'cysts' are found, and again these are found in any tissue but particularly in muscle, skeletal and cardiac, and in the brain; the 'cysts' are spherical and may reach a diameter of several hundred μm and contain masses of slowly multiplying trophozoites, referred to as bradyzoites (Fig. 118).

The final host of *T. gondii* is a member of the cat family, including domestic cats, and such infected hosts excrete oocysts in large

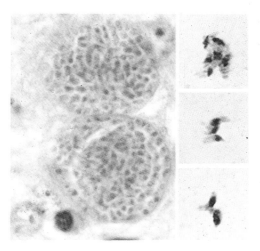

Fig. 118 *Toxoplasma gondii.* On the left is a section of mouse brain from an experimentally infected animal; two 'cysts' are shown and each contains numerous trophozoites. On the right are films made from the peritoneal exudate of the same mouse; free trophozoites are seen and these tend to show a crescent shape with a single central nucleus. Haematoxylin and Eosin.
Brain section $\times$ 1200; peritoneal exudate $\times$ 1200

numbers in their faeces. This offers another means by which humans may become infected. The oocyst is 10×12 μm, and once excreted sporulates to form eight sporozoites. The infective forms can survive for months under suitable conditions such as warm moist soil; thus, in addition to humans, other mammals and birds can become infected and man by virtue of his carnivorous nature can also become infected by eating inadequately cooked or raw meat of such other infected intermediate hosts.

Transplacental infection of the foetus, often from an asymptomatic mother, also occurs, with the prospect of abortion or the child being born with deformities.

LABORATORY DIAGNOSIS

Microscopic examination of smears and sections of biopsy material for trophozoites and large 'cysts' can be undertaken; similar such material can be inoculated into mice which are sacrificed six weeks later; the mouse brain is sectioned and examined for 'cysts'.

Sero-diagnostic tests are available, e.g. the Sabin-Feldman dye test, complement fixation, indirect haemagglutination and immunofluorescent techniques; much of the epidemiology of toxoplasmosis was elucidated by surveys using the dye test. Since the dye test involves the use of live *T. gondii* it is performed only in certain reference laboratories.

Trypanosomes

43

Trypanosomes are parasitic flagellated protozoa (Fig. 119) and those species which parasitise vertebrate hosts require a vector for transmission between hosts. Mammalian trypanosomes are either *stercorarian* or *salivarian* depending on the location in the insect vector of the metacyclic forms, i.e. the infective forms which appear at the end of the life cycle in the insect; stercorarian trypanosomes are those in which the metacyclic forms occupy a posterior site in the gut and are excreted in the insect's faeces. The metacyclic forms of salivarian trypanosomes are in the insect's proboscis and are injected into the host when the vector feeds.

Although most mammalian trypanosomes are stercorarian they display a high degree of host specificity and are not pathogenic but one species, *Trypanosoma cruzi* is exceptional in being pathogenic and for a wide variety of hosts.

TRYPANOSOMA CRUZI

This species is widely distributed throughout South America and the southern states of the USA. In addition to man, in whom it causes Chagas' disease (Fig. 120), it parasitises a wide range of wild and domestic animals. The vectors are triatomid bugs — cone-nosed bugs, and Chagas' disease is frequently acquired by those living in bug-infested dwellings; the disease is frequently zoonotic in origin but human-to-human infection via the vector does occur.

Metacyclic trypanosomes are deposited on the skin of the host since the bug usually defaecates whilst it is feeding, and the infective forms may enter the host via the puncture wound; however, since the bugs feed at night and often on the host's face the metacyclic forms can be spread over the face and enter the host's conjunctivae, nose or mouth.

The *trypomastigote* form then circulates in the bloodstream before invading tissue cells, particularly muscle; in the tissues the trypanosome changes to the *amastigote* form and reproduces by binary fission with the formation of a 'colony' containing many hundreds of amastigotes (Fig. 121).

After reproduction in the tissues the cells revert to the trypomastigote form and again invade the bloodstream; thus alternating cycles of reproduction in the tissues and invasion of the bloodstream occur. As infection continues the trypomastigotes become increasingly rare.

When a cone-nosed bug feeds on a patient who has trypomastigotes circulating in his blood, the trypomastigotes then develop in the insect's midgut and *epimastigote* forms

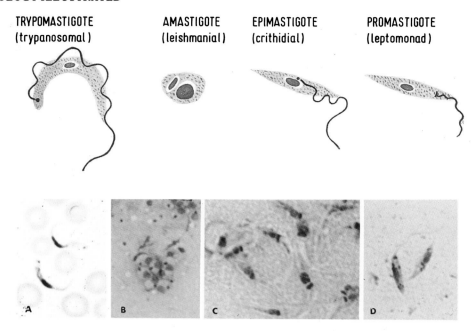

| TRYPOMASTIGOTE (trypanosomal) | AMASTIGOTE (leishmanial) | EPIMASTIGOTE (crithidial) | PROMASTIGOTE (leptomonad) |

Fig. 119 Trypanosomes do not reproduce sexually and commonly multiply by binary fission; they are pleomorphic in that the form of the cell varies during the life cycle and the following terms describe the different forms

Trypomastigote (trypanosomal form) when the kinetoplast is situated posterior to the nucleus
Amastigote (leishmanial form) does not possess a flagellum
Epimastigote (crithidial form) when the kinetoplast is adjacent to the nucleus
Promastigote (leptomonad form) in which the kinetoplast is near the anterior of the cell
Stained preparations from blood films (A–D) are shown below the respective diagrammatic representations

appear which spread to the hind gut and within a week or so metacyclic forms are excreted by the bug.

Laboratory diagnosis

This can be attempted by microscopic examination of blood films taken during the early, acute stage of infection; *T. cruzi* in the trypomastigote form in the blood has a characteristic appearance, some 20 μm in length and with a curved, often C-shaped, appearance (Fig. 122). The centrally placed nucleus stains darkly and an even more darkly stained kinetoplast lies in the posterior part of the cell; the undulating membrane associated with the flagellum is often seen. The flagellum arises from the vicinity of the kinetoplast and passes along the cell surface to point anteriorly.

Xenodiagnosis may also be undertaken; i.e. laboratory-reared bugs are allowed to feed on the suspected case and after some days the bugs' faeces are examined for trypanosomes.

In chronic infection serological diagnosis must be relied on and, although complement-fixation tests with patient's serum have been widely used in the past, haemagglutination and fluorescent antibody techniques are now available and are much more sensitive.

Of the several non-pathogenic stercorarian trypanosomes, only *T. rangeli* is likely to cause confusion in the laboratory diagnosis of Chagas' disease since it has the same host range and vectors as *T. cruzi* and occurs in the same parts of the world; the two species can be easily differentiated in blood films after a little experience.

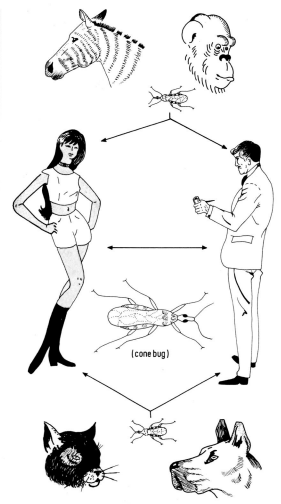

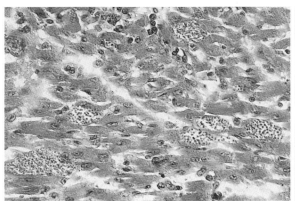

(cone bug)

Fig. 120 Chagas' disease. This infection is caused by *Trypanosoma cruzi* which parasitises a wide range of wild and domestic animals, whence it is transmitted to man by triatomid (cone-nosed) bugs; human-to-human infection by the vector can also take place

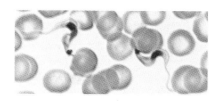

Fig. 122 *Trypanosoma cruzi.* The characteristic curved shape is evident in this blood film, as is the darkly stained posteriorly sited kinetoplast. Thick blood films are more satisfactory for diagnosis, since in thin films the trypanosomes frequently break up with loss of much of the cytoplasm
× 1200

TRYPANOSOMA GAMBIENSE

This salivarian trypanosome is primarily a human parasite in West and Central Africa within the latitudes 15°N and 30°S; transmission from person to person is by the bite of tse-tse flies of the *Glossina palpalis* group, and *T. gambiense* causes a chronic form of sleeping sickness lasting months or years.

In the infected patient the trypanosomes reproduce by longitudinal binary fission in the blood and other body fluids and recurring parasitaemia is a feature of the early stages of the illness; the organisms can also be detected in enlarged lymph glands, and later in the disease, when trypanosomes are scanty or absent from the bloodstream, they can be found in the cerebrospinal fluid.

Development of either species within the tse-tse fly is identical but much more complex than in the human host; trypanosomes ingested by the fly pass to the midgut where within the peritrophic membrane they assume

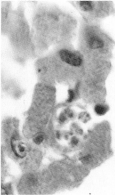

Fig. 121 *Trypanosoma cruzi.* Sections of heart muscle of mouse infected with *T. cruzi.* On the left (× 250) at least nine distinct clusters of parasites, embedded in the muscle fibres, can be seen as collections of blue dots; on the right (× 1000) and immediately below centre the nuclei and kinetoplasts of the amastigote form of *T. cruzi* can be seen. Haematoxylin and Eosin

187

a long broad form. They escape at the posterior end of the midgut to lie between the membrane and the gut wall and a week or ten days after being ingested the parasites are found closely packed in the extra-peritrophic space. Here they have become long and slender; they must then penetrate the peritrophic membrane and reach the main food-channel to migrate along the foregut to the proboscis. Thereafter they pass up the hypopharynx to the salivary glands where they change into epimastigote forms; metacyclic trypanosomes develop from the epimastigotes and pass with the saliva down the hypopharynx to be inoculated when the fly bites a new host. Throughout the development in the tse-tse fly reproduction takes place by binary fission.

TRYPANOSOMA RHODESIENSE

Infection with this salivarian protozoan is restricted to East Africa and it affects many wild and domestic ungulates including cattle, which are the primary hosts; man is incidentally involved when bitten by tse-tse fly vectors such as *Glossina morsitans* or *G. pallipides;* the acute form of sleeping sickness which results is therefore a zoonotic infection.

Development within the vector fly is identical to that of *T. gambiense.*

LABORATORY DIAGNOSIS (Fig. 123)

This depends primarily on the microscopic detection of the parasite in blood films, aspirated lymph gland juice or cerebrospinal fluid. Inoculation of such material intraperitoneally into mice or rats is also an efficient detector of *T. rhodesiense* (these animals are not susceptible to infection with *T. gambiense);* tail blood from the inoculated animal is examined for the presence of trypanosomes from the third day onward.

Sero-diagnostic procedures are not presently useful because of the antigenic variability of trypanosomes.

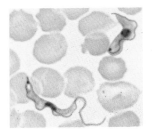

Fig. 123 *Trypanosoma gambiense.* Morphologically *T. gambiense* is identical with *T. rhodesiense* and *T. brucei.* The pleomorphic nature of *T. gambiense* is exemplified by comparing the stumpy form (top left), which has no free flagellum, with the slender form. The small sub-terminal kinetoplast is not as conspicuous as in *T. cruzi*
× 1200

OTHER TRYPANOSOMES

Many other trypanosomes are associated with diseases in animals, e.g. *T. brucei,* which has a wide range of hosts some of which suffer severe and frequently fatal infections whereas others display little or no effect from the parasitisation; with the exception of infection with *T. equiperdum,* which is transmitted between horse or donkey hosts by sexual intercourse, all of these animal trypanosomes require an insect vector.

LEISHMANIAE

Four species of *Leishmania* parasitise man and other mammals, but only man suffers clinical infection; transmission is by sandflies, and, although human-to-human infection is common in certain geographical areas, zoonotic spread, e.g. from dogs, foxes or jackals, also occurs. Such animals become infected when eating infected carcasses, and this forms an additional method of maintaining an animal reservoir of leishmaniae.

The four species of *Leishmania* are morphologically identical and they are separated essentially by the type of clinical illness produced in the human host;

leishmaniae differ from trypanosomes in that their development in the mammalian host is solely intracellular.

LEISHMANIA DONOVANI

L. donovani causes visceral leishmaniasis, commonly known by its Indian name, kala-azar; transmission by infected sandflies from human to human is common, but dogs also act as a fruitful source of infection via the same vector.

L. donovani introduced to a host by the sandfly's bite first proliferates at the site of inoculation and then spreads by the lymphatics and bloodstream to all organs possessing reticulo-endothelial cells; in the host the promastigote forms introduced by the sandfly rapidly change to the amastigote form and are ingested by macrophages within which they divide by binary fission until the host cell is crammed with parasites. When the host cell ruptures the amastigote forms are liberated to be taken up by fresh macrophages.

If a sandfly takes a blood meal from a host when infected macrophages are available, some of these will be ingested, and the

amastigotes are released from the macrophages in the insect's midgut and change to the promastigote forms; these multiply rapidly and migrate to the foregut ready to be inoculated into a fresh host; from ingestion of infected macrophages by the sandfly an interval of 5–7 days elapses before promastigotes appear in the foregut.

The amastigote form which infects man is an ovoid cell, some 2–4 μm in size, and does not possess flagella; each cell possesses a nucleus and a kinetoplast. Cells in this form are referred to as Leishman-Donovan (LD) bodies and are diagnostic of leishmaniasis (Fig. 124).

LEISHMANIA TROPICA

This species is identical with *L. donovani* in morphology and in its development both in the insect vector and mammalian hosts; it parasitises white blood cells but causes lesions only in the skin (cutaneous leishmaniasis). Although popular names for the condition are oriental sore or Delhi boil, the infection also occurs in Mediterranean countries and in areas on the north and west coasts of Africa.

LEISHMANIA BRAZILIENSIS

Like *L. tropica,* this species invades the skin and also mucous surfaces (muco-cutaneous leishmaniasis); metastatic spread occurs and in South America the disease, espundia, affects the mucosa of the nasopharynx and also the nasal cartilage.

The fourth species is named *Leishmania mexicana* and produces lesions similar to those of *L. braziliensis.*

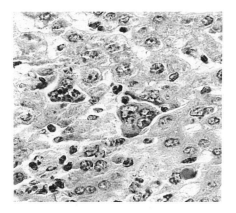

Fig. 124 *Leishmania donovani.* Section of spleen showing Leishman-Donovan bodies; these comprise numerous amastigote forms of the parasite within the host cells.
× 500 Haematoxylin and Eosin

LABORATORY DIAGNOSIS

In visceral leishmaniasis, smears should be made from bone marrow, spleen and liver

189

aspirates, and after staining a search made for the amastigotes, i.e. Leishman-Donovan bodies. Likewise, aspirated material can be inoculated into NNN medium which is then incubated at 25°C (incubation at 37°C is lethal to the parasite) and although growth is slow, cultivation often succeeds in confirming the diagnosis when the parasites are too scanty to be recognised microscopically in smears. In culture the protozoon develops only to the promastigote stage.

Hamsters are susceptible to infection, and material from the patient should be inoculated intradermally or intraperitoneally; skin and spleen biopsies are examined for Leishman-Donovan bodies some weeks later.

Similar procedures are undertaken in cases of cutaneous and muco-cutaneous leishmaniasis, using material acquired by needle aspiration or curettage of the lesions.

Serological diagnosis, using an indirect fluorescent antibody technique, is available in certain Reference Laboratories.

Malaria

44

Man is the intermediate host for at least four species of plasmodia, *Plasmodium falciparum*, *P. malariae*, *P. ovale* and *P. vivax*; two other species, *P. cynomolgi* and *P. knowlesii*, whose natural intermediate hosts are simians, can also be transmitted to man at least experimentally.

The definitive hosts for these plasmodia are various species of anopheline mosquito which do not suffer any ill effect from parasitisation; contrarily man suffers malaria which, in the absence of specific therapy, is characterised by regularly recurring attacks of severe fever. Malaria occurs extensively throughout those parts of the world where conditions are suitable for breeding of the relevant mosquito and where they have access to the human host.

Malaria is one of the most important infections in the world today, both in regard to morbidity and mortality, and its effect on the economy of countries such as India must be tremendous, although control measures have effected its eradication from certain areas.

Determined and exhaustive malaria eradication schemes were begun under the auspices of the World Health Organisation in the 1960s but have largely failed for the following reasons:

1. Not all dwellings were treated with insecticides to kill the vector.
2. Mosquitoes acquired resistance to insecticides.
3. In several areas there was resistance to the eradication scheme, notably amongst the wealthier members of the population.
4. The schemes were disrupted by wars and other political upheavals associated with a shortage of money.

Persons normally resident in temperate climates where malaria does not occur naturally have experienced a dramatic change in regard to malarial infection by importation; usually imported cases are associated with migrants to Britain who have returned from a short holiday to their tropical country of origin and have been negligent in their use of prophylactic drugs. Native Britons also contribute to the increase in imported cases of malaria, and it is not sufficiently appreciated that even a brief stop-over in an area where malaria is endemic can allow infection to take place. It should be mandatory in history-taking, particularly in cases of pyrexia of uncertain origin, to ask specifically regarding foreign travel and chemoprophylaxis, bearing in mind that plasmodia are developing resistance to antimalarial drugs; each year in Britain deaths do occur from malaria!

The complete life history of malarial parasites was elucidated little more than 30 years ago when experiments with monkey hosts revealed that injected parasites migrated rapidly to the parenchyma cells of the liver; subsequently a human volunteer was infected and subsequent liver biopsy showed that a pre-erythrocytic cycle also occurs in man.

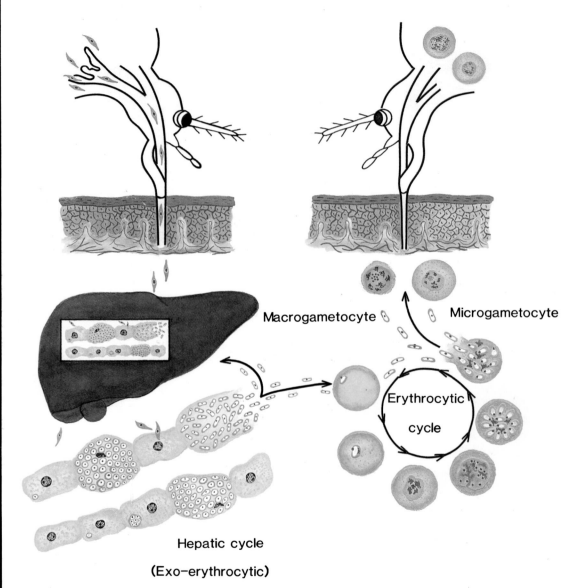

Macrogametocyte

Microgametocyte

Erythrocytic cycle

Hepatic cycle

(Exo-erythrocytic)

Fig. 125 Schizogony. This diagram summarises the sequence of events from the time that sporozoites are inoculated from the mosquito on the left into man, through the exo-erythrocytic hepatic cycle, then the erythrocytic cycle until another mosquito (right) takes a blood meal from the patient and ingests gametocytes along with merozoites. The 'window' in the liver is shown in enlarged form immediately below that organ and depicts the development of schizonts and the release of merozoites when the liver cells rupture; similarly the progression of the erythrocytic cycle from the invading ring form through the formation of trophozoites to schizonts is shown. The merozoites liberated from erythrocytic schizonts can then invade fresh red blood cells, healthy liver parenchyma cells or be ingested by another mosquito along with gametocytes

That part of the life cycle of malaria plasmodia which occurs in the human host is known as schizogony and is asexual, although gametocytes are formed in the host; the sexual cycle, or sporogony, takes place in the female anopheline mosquito.

SCHIZOGONY (Fig. 125) see opposite

After the human host has been bitten by an infected female mosquito, the sporozoites thus introduced leave the general circulation within a short time and invade tissue cells with a predilection for the parenchyma cells of the liver; within these latter cells the sporozoites divide asexually to form schizonts which ultimately occupy the entire liver cell and are teeming with thousands of merozoites. At the end of this pre-erythrocytic cycle, which lasts 7–10 days, the infected liver cells rupture and the liberated merozoites invade red blood cells in the general circulation.

This pre-erythrocytic cycle is of variable duration depending on the species of plasmodium causing the infection; merozoites from *P. vivax* and *P. ovale* infection can also invade fresh liver cells directly so that an exo-erythrocytic cycle continues in the liver in parallel with the erythrocytic cycle. Schizogony is thus a continuing process in the liver and red blood cells, and except in infection with *P. falciparum* and *P. malariae* the persistence of an exo-erythrocytic hepatic cycle gives rise to relapses of the infection. On occasion the time interval between the victim being bitten by an infected mosquito and the appearance of plasmodia in the circulating blood may be grossly prolonged, even for many months; in such cases it is thought that the parasite, having entered the liver cells, becomes dormant for some reason, and the term 'hypnozoite' has been coined for this parasitic phase. The factors involved in such dormancy are a matter for speculation.

The duration of the erythrocytic cycle also varies depending on the parasite involved and is determined by the interval required for the merozoite infecting a red cell to pass from the ring-form stage to become a trophozoite and in turn a schizont, with the ultimate rupture of the red cell and the liberation of new merozoites; the interval is 48 h in the case of *P. ovale* and *P. vivax*, 36–48 h for *P. falciparum* and 72 h for *P. malariae*.

Thus classically, in a human host bitten on only one occasion and infected with only one species, schizogony will be synchronous since virtually all the parasites will mature simultaneously, with massive numbers of merozoites being liberated in a short space of time, so that febrile paroxysms will occur every second day in the case of infection with *P. vivax* or *P. ovale* and every third day with *P. malariae* infections. This periodicity is influenced, however, not only by therapy but also if the patient suffers multiple infection with the same species or if more than one species infects the same host.

During the erythrocytic cycle a few parasites become differentiated as either male (microgametocyte) or female (macrogametocyte) cells so that the sexual cycle commences in the human host.

SPOROGONY (Fig. 126)

Before sporogony can be fulfilled, the male and female gametocytes must be ingested by a female mosquito, and further development of the gametocytes takes place in the stomach or midgut of the vector once the gametocytes have escaped from the ingested red cells. The macrogametocyte matures to become a macrogamete by forming one or two polar bodies which become detached; the macrogamete is then ready for fertilisation.

Maturation of the microgametocyte is known as exflagellation, and is a somewhat more complex process; firstly the nucleus of the microgametocyte divides three times and the resultant eight nuclei move to the periphery of the cell, and each then moves into separate flagellar-like structures which have formed at the cell surface; the microgamete is then ready to fertilise a

macrogamete with the formation of a zygote. The zygote in turn becomes an ookinete, a crescent-shaped motile cell which migrates through the gut epithelium within 24 hours or

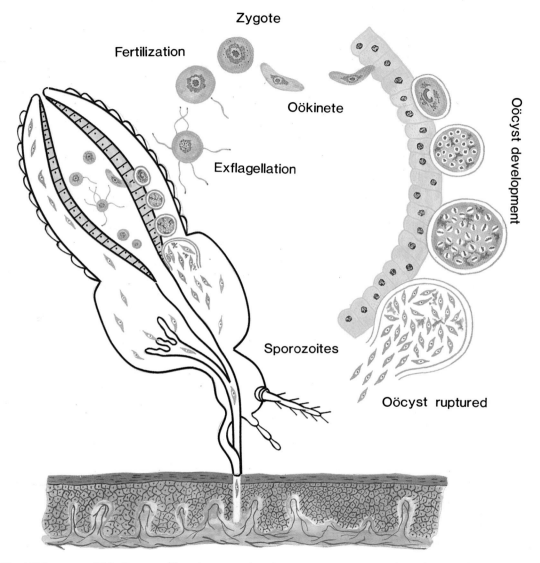

Zygote

Fertilization

Oökinete

Exflagellation

Oöcyst development

Sporozoites

Oöcyst ruptured

Fig. 126 Sporogony. This diagram outlines the progression of events in the sexual stage of the life-cycle of malaria parasites in the mosquito; two gametocytes have been ingested and after maturing to macrogamete and microgamete — the latter showing exflagellation, fertilization occurs with the formation of a zygote which develops into the motile, crescent-shaped ookinete which then penetrates the gut wall to lie immediately under the basement membrane where the oocyst forms with the eventual liberation of sporozoites into the coelom. The sporozoites migrate to the salivary glands and await injection into a human host when the mosquito next takes a blood meal

so of being formed. Further development takes place on the outer surface of the stomach under the basement membrane; an oocyst is formed by rounding up of the ookinete and the development of an encysting membrane, and within the growing oocyst the nucleus divides repeatedly and ultimately the cytoplasm divides to form hundreds of sporozoites. These infective forms of the parasite are liberated into the coelom whence they migrate to the salivary glands of the mosquito to await injection into another host.

LABORATORY DIAGNOSIS

This is based on the microscopic examination of stained blood films; thick films are used to scan for the presence of malarial parasites and identification of the species infecting the patient is based on examination of thin films.

Species identification rests on the differing morphology of the erythrocytic forms of the parasites, and some of these are exemplified photographically (Figs. 127–130).

A thin blood film is prepared by collecting

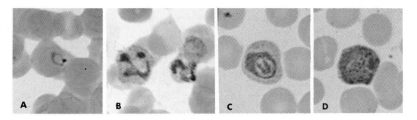

Fig. 127 Plasmodium vivax. At **A** the merozoite has assumed a ring form soon after entering a red cell; a rim of cytoplasm surrounds the central vacuole and the nucleus lies within the cytoplasm — the nucleus is referred to as the chromatin dot.

In **B** are developing trophozoites which arise from the ring form within a few hours; the vacuoles are reduced in size and the parasites are irregularly shaped.

C shows a microgametocyte with the nucleus centrally placed and comprising a diffuse skein of fibrils.

D A macrogametocyte; this female form is larger than the microgametocyte and expands the red cell. The nucleus is characteristically compact and sited peripherally, in this instance at the right side of the parasite.

× 1200

It should be noted that the gametocytes of *P. malariae* and *P. ovale* are similar to those of *P. vivax* in their general characteristics but are not so frequently seen in peripheral blood. Those of *P. malariae* are smaller than *P. vivax* whereas the gametocytes of *P. ovale* are intermediate in size

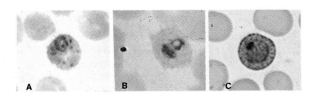

Fig. 128 Plasmodium ovale. **A.** A ring form which is thicker than the delicate ring form of *P. vivax* **B.** A compact trophozoite with a large nucleus; deformity of the parasitised red cell is characteristic of *P. ovale* infection **C.** Another compact trophozoite; the multitude of Schuffner's dots almost obscure the parasite

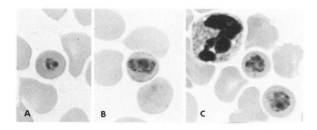

Fig. 129 Plasmodium malariae. **A.** shows a ring form and at **B.** a compact band form of the developing trophozoite; this form is common in *P. malariae* infections.

Two immature schizonts are seen at **C** — centre and bottom right.

× 1200

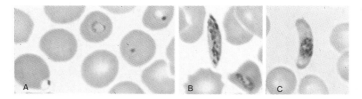

Fig. 130 Plasmodium falciparum. **A.** Ring forms with conspicuous chromatin dots; at lower left the ring form is marginally situated and at top centre the chromatin dot is double. **B.** The microgametocyte shows diffuse scattering of chromatin granules and pigment. **C.** In comparison with **B** this macrogametocyte is characteristically more curved and has a compact mass of chromatin and pigment near the centre

on the narrow edge of a microscope slide a drop of blood about the size of a pin head; this spreader slide is then applied at an angle of 30° to a second slide on a horizontal surface and pushed along so that the drop of *blood is trailed behind* the interface of the slides; if the blood is pushed instead of being dragged then the malarial parasites may be crushed.

A thick blood film is prepared by collecting in the centre of a slide a drop of blood two or three times the volume of that required in preparing a thin film; the drop is then spread in a circle 1–2 cm in diameter using the corner of another slide. The films are dried rapidly and the thin film stained by Leishman's or Giemsa's methods, which allow visualisation of plasmodia within intact red cells; parasitised r.b.c.s tend to roll to the edges of the blood film, and also to be carried to the tails of the film, so that these areas should be examined first.

The thick film, perhaps 10–20 cell layers in depth, is opaque to transmitted light and must be made colourless by removing the haemoglobin; this is facilitated by using an aqueous stain, e.g. Field's, when staining and lysis occur at one and the same time. Stained thick films must *not* be blotted dry.

SECTION 5

Mycology

There are thousands of species of fungi. As with bacteria, the vast majority follow a saprophytic existence and, along with bacteria, assist in the destruction of plant and animal debris.

Fungi in their saprophytic role occasionally annoy microbiologists by contaminating laboratory culture media — and indeed such laboratory contamination with *Penicillium notatum* allowed Sir Alexander Fleming to note the antibacterial activity of this species; the species now used for the commercial production of penicillin is *Penicillium chrysogenum.* A few other useful antimicrobial agents had a similar origin.

Fungi can also indirectly irritate mankind by causing spoilage of foodstuffs and, more seriously, the production of fungal toxins — e.g. aflatoxin by *Aspergillus flavus* growing on maize — and their subsequent ingestion by man is incriminated as a cause of liver damage and perhaps even liver cancer. The evidence for such serious afflictions was first found in Thailand and certain African countries.

A very few fungi cause infection in man and animals, and additionally the spores of certain fungi, e.g. *Aspergillus clavatus,* are incriminated in allergic reactions, e.g. broncho-alveolitis.

One of the more recent impacts of fungal infections is the increasing incidence of life-threatening situations when even the most humble fungi are enabled to attack the patient whose defence mechanisms have been compromised by immunosuppressive therapy or whose normal bacterial flora has been dramatically altered by the administration of broad-spectrum antibacterial agents.

Classification and techniques in mycology

45

CLASSIFICATION OF FUNGI

Fungi pathogenic to man can be classified on a morphological basis into four groups:

Moulds which grow as branching filaments (hyphae) that interlace and form a dense, felted mass (mycelium); such filamentous fungi have a vegetative mycelium which grows into the substrate and absorbs nutrients whereas the aerial mycelium rises from the surface of the mould and allows dissemination of sexual spores.

Yeasts are unicellular fungi appearing as single round or oval cells; reproduction is by budding from the parent cell and sexual spores are not formed. When grown on solid media, yeasts form colonies similar to those of bacteria in contrast to the powdery colonies of moulds.

Yeast-like fungi grow either as round or ovoid cells or as non-branching filaments; like yeasts they also reproduce by budding and on solid media produce colonies similar to those of staphylococci.

Dimorphic fungi comprise the fourth morphological group and are so called since when growing in tissues or in vitro at 37°C they appear in yeast forms, whereas when cultures are incubated at 22°C they present as mycelial growth.

A more formal and systematic classification of fungi exists which is primarily dependent on the nature of their sexual processes; since these are difficult to induce the formal classification is not offered here but it should be noted that most of the fungi pathogenic to man fall into Class IV of that classification, i.e. the *Fungi imperfecti,* a class which groups together all fungi which do not have a sexual stage or where such a stage has not yet been demonstrated.

TECHNIQUES IN MYCOLOGY

Microscopy

Microscopic examination of unstained or stained material plays a large part in the identification of fungi; indeed microscopy of the specimen from the patient and from resultant cultures is frequently the only manoeuvre required to arrive at a diagnosis.

Specimens of skin scales, hair or nail clippings from cases of ringworm (tinea) must first be rendered transparent ('cleared') to allow observation of the infecting fungi; fragments of either type of specimen are placed on a glass slide and covered with 20% sodium or potassium hydroxide, and a cover slip is then applied. The preparation is then left at room temperature to allow the keratin to be partially dissolved; skin scales will be cleared within 5–10 min whereas pieces of nail may have to be treated for an hour or two at 37°C before digestion has reached the stage at which fungi may be seen. In that case the hydroxide must be replenished from time to time, or alternatively the specimen may be immersed in the agent in a test tube and incubated before being transferred to a microscope slide for examination.

The cover slip is pressed down gently with blotting paper to give a thin film which is then examined under the dry objectives; cholesterol crystals in skin, and oil or fat droplets, may be mistaken for fungal elements but this potential source of confusion can be eliminated by replacing the hydroxide with lactophenol blue stain which is not accepted by such artefacts.

Staining

If it is desired to stain a cleared preparation, lactophenol blue stain is applied at one edge of the cover slip and the hydroxide is drawn from the opposite edge with blotting paper until the stain has completely replaced the alkali (Fig. 131).

Gram's staining method is also used in the diagnosis of fungal infections, e.g. in candida infections, either directly on material from cases or on cultures obtained from specimens; India ink preparations may also be used, e.g. to demonstrate the capsules of *Cryptococcus neoformans* in cerebrospinal fluid.

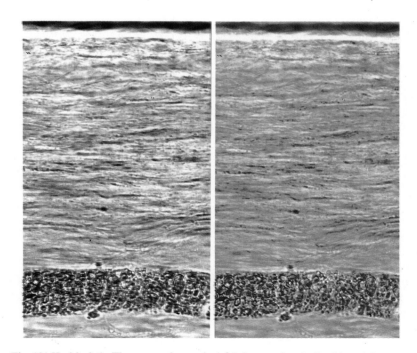

Fig. 131 *Healthy hair.* The preparation on the left is in potassium hydroxide and that on the right after the hydroxide was replaced with lactophenol blue; slightly more than half the width of the hair is shown with the surface cuticle at the top and the medullary core near the lower edge. The cortex comprises more than two thirds of the hair shaft. This preparation should be compared with those showing hairs infected with ringworm fungi
× 500

Cultivation

The most commonly used medium for diagnostic purposes is Sabouraud's glucose peptone agar, which has a pH of 5.4, and cultures may be made in parallel on this medium with or without the addition of thiamine (10 mg/l); the latter promotes spore formation by certain fungi causing ringworm. The aerobic nature of fungi must be catered for when attempting their isolation, as must the fact that they grow more slowly than bacteria and culture plates should be incubated at 20°C for at least three weeks before being discarded although most positive cultures will show evidence of growth within a few days and be typical for identification by 7–14 days.

In the case of ringworm fungi colonial appearances should be noted not only of the surface growth but also on the reverse side of the colony and attention should be paid to any pigment which may have diffused into the medium; the majority of ringworm fungi give large spreading colonies covered by a fluffy or powdery aerial mycelium; the various pigments which are produced may be confined to the colony or diffuse into the medium for a greater or lesser distance. Some colonies are exemplified but no attempt has been made to give any detail of colonial morphology.

Needle mount preparations

Microscopic examination of material from colonies allows differentiation of the three genera of fungi responsible for ringworm; a needle mount preparation is obtained by removing a piece of sporing mycelium from a culture on Sabouraud's agar; the material is placed on a slide and is then teased out in a drop of 95% ethyl alcohol using two straight wires or needles. Just before the alcohol has completely evaporated a drop of lactophenol blue stain is added and a cover slip is applied; the preparation is then left at room temperature for a few minutes to allow the stain to penetrate. Excess stain is then removed by gentle pressure on the preparation through a sheet of blotting paper before viewing with low-power and high-power dry objectives.

Slide culture method (Fig. 132)

By subculturing mycelial growth on an agar block held between a microscope slide and a cover slip the arrangement of the growing mycelium and spores can be observed intermittently and undisturbed during growth.

A block of Sabouraud's agar medium, 1 cm square and 2 mm thick is placed on a sterile microscope slide and the block covered with a sterile cover slip; using a straight wire the agar block is inoculated at the midpoint of each of the vertical sides with material from a culture. The assembly is then placed in a closed chamber, e.g. a plastic sandwich box, containing several layers of blotting paper

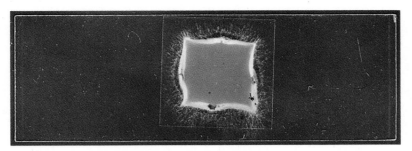

Fig. 132 Slide culture technique. A block of sterile Sabouraud agar is held between the glass slide and cover slip; the medium has been inoculated on each of its free vertical sides with an *Aspergillus* species and naked eye growth is clearly seen following suitable incubation

soaked in 20% glycerol to ensure a humid aerated atmosphere. Incubation is at room temperature, or preferably at 28°C.

The slide can be removed at intervals and examined microscopically with the dry objectives without disturbing the assembly; as soon as adequate sporing is noted, stained preparations can be made from both the slide and the cover slip.

The cover slip is removed and the agar block discarded; one drop of 95% ethyl alcohol is applied to the slide and to the cover slip and immediately before the alcohol has completely evaporated one drop of lactophenol blue stain is applied to each preparation; these are then covered with a cover slip and slide respectively. The preparations are left at room temperature for some time to allow penetration of the stain before the excess is removed by gentle pressure through a sheet of blotting paper; if permananent preparations are required the edges of the cover slip can be sealed to the slide, e.g. with nail varnish.

Dermatophytic fungi

46

Three genera of the Fungi imperfecti, namely *Microsporum, Trichophyton* and *Epidermophyton,* provide the species causing the most common type of fungi infection throughout the world without discrimination between sexes, race or climate; the dermatophytoses, commonly called tinea or ringworm, are superficial infections of the skin, hair and nails which never spread deeper than the keratinous layer but spread peripherally from foci to produce ring-like lesions. Species of ringworm fungi which primarily infect man are *anthropophilic*, and these usually produce chronic mild and often subclinical infections, although scalp infection with *T. schoenleinii* is exceptional and causes favus; this infection is sited in hair follicles and gives rise to small red papules which develop into scutula, i.e. cup-shaped, sulphur-yellow crusts which eventually destroy the affected hairs.

Among the anthropophilic species some, e.g. *Trichophyton mentagrophytes* and *T. tonsurans*, are catholic in their habits and can attack nails and the skin and hair anywhere on the body, whereas other species are more restricted in the sites where they cause disease, e.g. *Epidermophyton floccosum*, which is most commonly incriminated in tinea cruris and can also infect nails and other skin areas but does not cause infection of the scalp.

Similarly *Microsporum* species, commonly the cause of scalp ringworm, do not affect nails. *M. audouinii* infections are restricted to children, in whom it is the commonest cause of scalp ringworm, and such infection usually disappears spontaneously around puberty.

On the other hand *zoophilic* species are parasitic essentially on various animals but can be transmitted to human hosts, especially children, from domestic pets or farm animals; when such zoophilic species as *M. canis, M. gypseum* and *T. equinum* cause human infection they usually cause a more acute inflammatory response than do anthropophilic species.

Laboratory diagnosis of ringworm

As with specimens for bacteriological examination, material submitted for laboratory diagnosis must be taken carefully. Cleansing of the lesion with 70% alcohol will reduce the chance of bacterial contamination; skin scales should be harvested by scraping the skin at the advancing edge of the lesion with a blunt scalpel. Infected nails are clipped for examination and scrapings are also taken from the deeper parts using a blunt scapel. Great care is required in selecting hairs for submission to the laboratory; certain species cause the infected hairs to fluoresce under

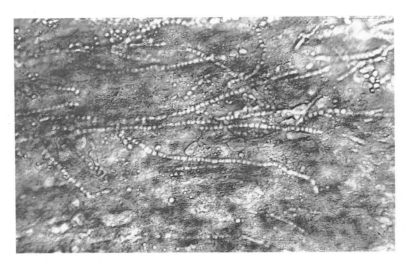

Fig. 133 *Trichophyton tonsurans.* Hair from an early case of tinea capitis in a child; chains of large arthrospores are visible in this endothrix infection
 × 500 KOH preparation

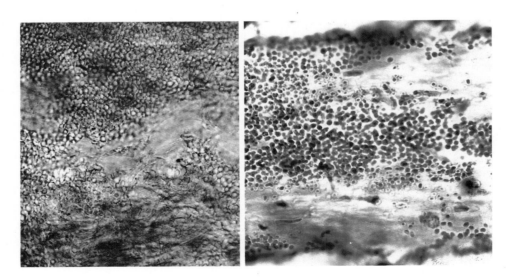

Fig. 134 *Microsporum audouinii.* Hairs from a case of tinea capitis in a child; the preparation on the left is in potassium hydroxide and that on the right after the hydroxide was replaced with lactophenol blue. This is a classical example of small-spore ectothrix ringworm infection of hair
 × 500

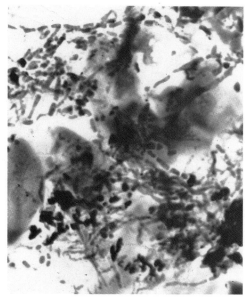

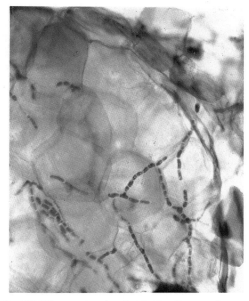

Fig. 135 Film of material from a scutulum in a case of favus due to *T. schoenleinii,* showing a dense felt work of mycelium and spores with a background of cellular debris
    × 500 Lactophenol blue stain

Fig. 136 Skin scraping showing hyphae fragmenting to form arthrospores which, either by direct contact or indirectly by fomites, e.g., duck boards in spray rooms, can be transmitted to fresh human hosts
    × 500 Lactophenol blue stain

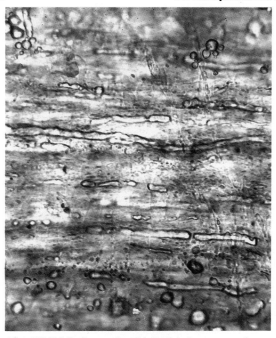

Fig. 137 *Trichophyton schoenleinii.* Hair from a case of favus showing endothrix infection: a few hyphae can be seen and also empty areas or tunnels where hyphae have degenerated. The presence of airbubbles is distinctive and the absence of the medulla of healthy hair should be noted
    × 500 KOH preparation

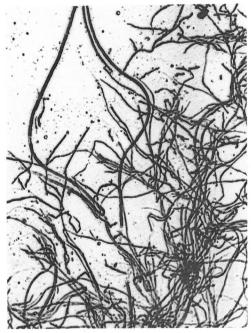

Fig. 138 *Microsporum audouinii.* Needle mount preparation from culture on Sabouraud's medium; the virtual absence of micro- and macroconidia is characteristic of this species when grown on Sabouraud's medium, and the growth here comprises only sterile interlacing hyphae

    × 500 Lactophenol blue stain

Fig. 139 *Microsporum canis.* Needle mount preparation from culture on Sabouraud's medium showing a cluster of large, thick-walled, fusiform macroconidia which are characteristic of most *Microsporum* species. In these mature macroconidia transverse septa have divided each into numerous cells or segments

    × 500 Lactophenol blue stain

Fig. 140 *Epidermophytonfloccosum.* Needle mount preparation from a young culture on Sabouraud's medium showing numerous thin-walled macroconidia which are beginning to assume a club-shaped appearance; the majority are non-septate but a few, e.g., at bottom right corner, show the characteristic formation of 2–4 cells. Microconidia are characteristically absent

    × 500 Lactophenol blue stain

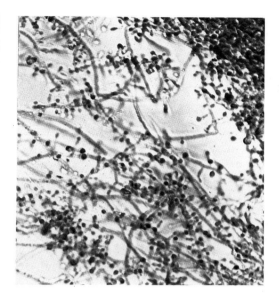

Fig. 141 *Trichophyton mentagrophytes*. Needle mount preparation from culture on Sabouraud's medium showing numerous microconidia; the majority are spherical and borne single from the sides of the hyphae. Macroconidia are absent as is the case with most *Trichophyton* species when grown on this medium
× 500 Lactophenol blue stain

Fig. 142 *Trichophyton mentagrophytes*. This needle mount preparation is exceptional since macroconidia are present in the growth from Sabouraud's medium; six cylindrical, thin-walled macroconidia are present and two of these show septa with division of the macroconidia into segments
× 500 Lactophenol blue stain

ultra-violet irradiation; for this purpose a Wood's lamp is used to view the head in a darkened room. The stumps of broken hairs should be removed with fine forceps and lustreless hairs from the vicinity of the lesion can also be sent for examination. All specimens are submitted in individual paper envelopes clearly marked for identification of the source.

Laboratory diagnosis is normally by microscopic examination (Figs 133–137) of the infected material after it has been cleared with alkali; microscopic examination of affected hairs shows whether infection is endothrix, as in the case of favus, or ectothrix; a note should also be made in the case of ectothrix infections whether infection is of the small-spore or large-spore type. Characteristically *M. audouinii* gives small-spore ectothrix infection as does *T. mentagrophytes*; other *Trichophyton* species produce large-spore ectothrix or alternatively endothrix infections.

Cultivation

Since bacteria may also be present on the specimen it is necessary to treat it with 70% alcohol for 2–3 minutes to reduce the bacterial population before inoculating media. The dermatophytic fungi are aerobic so that fragments of hair, skin or nail must be implanted on the surface and partially submerged. If a pure culture of the fungus is required from the resultant growth then subculture should be made from the *edge* of the spreading colony to fresh medium and this will usually avoid carrying over any bacterial contaminants which may be present.

Needle mount preparations from the aerial mycelium, when stained with lactophenol blue, reveal microscopically the size, shape and distribution of micro- and macroconidia (Figs. 138–142).

Microconidia are small, round or oval single-cell structures on spore-bearing hyphae, and are characteristically produced by *Trichophyton* species but are not seen in cultures of *Epidermophyton* species; they are seen only occasionally and in small numbers in *Microsporum*.

Similarly, the morphology of macroconidia can be noted and this further characterises the three genera of the dermatophytes:

Microsporum species produce large, thick-walled, fusiform macroconidia which are divided into numerous cells by transverse septa, although the anthropophilic species *M. audouinii* is exceptional in not producing conidia when grown on Sabouraud's medium.

Trichophyton species frequently do not yield macroconidia when grown on

Sabouraud's medium but when they occur they are small, thin-walled and cylindrical.

Epidermophyton species produce thin-walled macroconidia which in young cultures are finger-like and later assume a club or pear-shaped appearance.

The six plates, Figures 143–148, are intended to exemplify colonial appearances of some ringworm fungi.

All of the cultures were grown on Sabouraud's agar for 10 days at 20°C; in each instance the surface growth is shown on the left and the appearance of the base of the colony on the right.

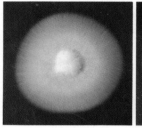
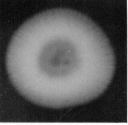

Fig. 143 *M. audouinii* (growth slow). Flat with sparse short aerial hyphae and creamish white in colour; the reverse is a light tan with a dark salmon-pink centre

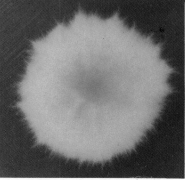

Fig. 144 *M. canis* (growth rapid). In this young culture the aerial mycelium is characteristically white and fluffy with yellow pigment showing through the peripheral growth; the reverse is characteristically yellow with a hint of the dull orange-brown of an older culture showing centrally

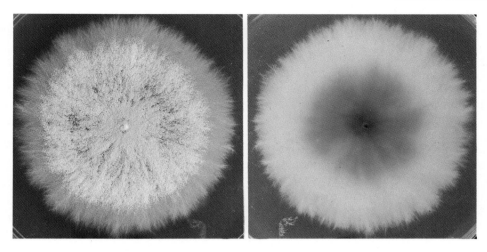

Fig. 145 *M. gypseum* (growth rapid). Flat with irregularly-fringed border and a powdery surface with a greyish-white centre and a cinnamon-brown edge. The reverse is tan with a red central area which is not often seen in this species

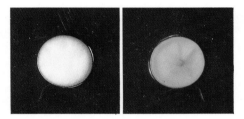

Fig. 146 *E. floccosum* (growth slow). The white, powdery, flat surface growth in this relatively young culture becomes tan or olive green later; the reverse is characteristically tan

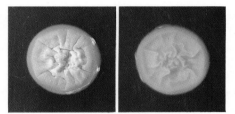

Fig. 147 *T. schoenleinii* (growth slow). The growth is characteristically irregularly heaped and folded with a waxy cream to yellowish-tan colour; the reverse is tan-coloured and the colony is somewhat reminescent of the surface of Camembert cheese

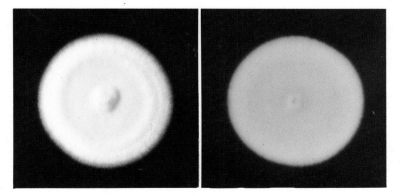

Fig. 148 *T. mentagrophytes* (growth slow). The downy, fluffy, flat surface growth is white at the periphery and light tan in the older central area; the reverse is tan in this instance but is often colourless and pink or even red

MALASSEZIA FURFUR

Malassezia furfur is the cause of a superficial skin infection known as pityriasis versicolor, which is worldwide in distribution although it is more commonly seen in hot humid climates; the infection is almost always asymptomatic and the chronicity of the infection, which produces irregularly shaped brownish desquamating lesions, dictates that it usually presents as a cosmetic problem. It must however be differentiated from more serious infections.

Skin scales from the lesion are examined microscopically and the presence of clusters of oval or round cells measuring 3–6 μm in diameter, together with short, thick hyphal elements approximately 2–4 μm in diameter, is sufficient to confirm the diagnosis (Fig. 149).

Cultivation of *M. furfur* from the skin scales is of little value, since similar organisms can be harvested from normal skin and differentiation of such organisms by biochemical and other methods is not yet valid.

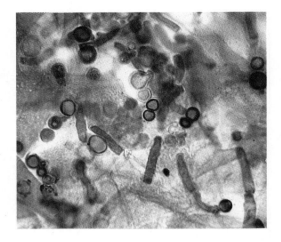

Fig. 149 *Malassezia furfur*. This preparation of skin scales from a case of pityriasis versicolor shows the characteristic round, thick-walled cells, 3–7 μ in diameter, and one of these (to the right of centre) is budding; also present are characteristic short, thick hyphal elements
×1000 Lactophenol blue stain

Candida albicans

47

Another fungal infection which is worldwide in distribution is candidosis (moniliasis) caused by *Candida albicans (Monilia albicans)* or more rarely by other species within the genus which are usually commensal in man and animals. Infection is usually confined to superficial lesions of mucous membranes, particularly in the mouth or vagina, or on the skin.

C. albicans spreads usually by direct contact with colonised individuals and is not a significant part of the microbial population in the environment, even in the vicinity of infected patients.

C. albicans is an opportunist pathogen and the host defence mechanisms have to be weakened by some other process before it can cause infection; oral candidosis (oral thrush) is most often seen in debilitated new-born babies and vaginal thrush occurs in diabetic women and also not uncommonly in pregnancy when it clears spontaneously after delivery. Vaginal carrier rates of 15% and 30% respectively are found in non-pregnant and pregnant women; these rates have increased to such levels since the introduction of oral contraception.

Skin infection (cutaneous candidosis) is usually restricted to flexures and other moist areas of the body but can also occur on the hands of people whose occupation demands prolonged immersion of their hands in water. Diabetics, alcoholics and drug addicts are peculiarly susceptible to infection, as are people suffering from leukaemia or iron-deficiency anaemia; similarly infection is promoted if a debilitated host is treated with cortisone or with broad spectrum antibiotics; *C. albicans* is resistant to most antibiotics.

Deep-seated infection, e.g. of the lungs and the intestine, can also occur as can generalised candidosis, and these serious infections are usually associated with the administration of antibiotics for the treatment of bacterial infections or with the use of immuno-suppressive agents.

LABORATORY DIAGNOSIS

Since *C. albicans* occurs as a commensal at sites where it may declare its opportunistic pathogenic role, the diagnosis of candidiasis in these sites is essentially a clinical one. However, the finding of large numbers of candida in a specimen, especially on repeated occasions, combined with the absence of other pathogens, assists in confirming the diagnosis.

It should be noted however that *C. albicans* is *never* present, in health, in the blood stream or in tissues. Thus its presence in such specimens is equated with infection.

211

Microscopy

Material from patches of thrush should be examined after being stained by Gram's method; budding yeast cells mixed with long filaments, pseudohyphae, are typical and the contrast in size with any bacteria which may be present leaves no doubt as to the fungal nature of the large Gram-positive yeast cells.

Cultivation

The specimen should be plated out on Sabouraud's medium and specimens of sputum or faeces from cases of suspected deep candidosis should also be inoculated on to a malt tellurite agar and a penicillin-streptomycin blood agar, since these selective media help to control the growth of bacterial species which will also be present.

    C. albicans gives large, cream-coloured colonies after 2–3 days incubation but cannot readily be differentiated from other species of *Candida* although colonial appearances on differential media, e.g. tetrazolium or eosin-methylene blue, can provide a presumptive identification and allow the selection of colonies for further study.

    In the diagnostic laboratory, confirmation of species identity is usually obtained by subculture into a series of sugars and noting the pattern of acid and gas production, but it is known that the fermentation reactions of *Candida* species are not constant.

    Other more specific tests to identify *C. albicans* are available and one of these, the pseudo-germ tube test (Fig. 150), deserves wider use; the test is simple, rapid and reliable and requires only the inoculation of a predominantly yeast-forming culture of *C. albicans* into undiluted human or animal serum. The mixture is examined microscopically after 2 h incubation at 37°C for the presence of pseudo-germ tubes emerging from the yeast cells.

    There is an increasing awareness that species other than *C. albicans* can cause not only oral and vaginal thrush but are also incriminated in deep-seated infections; of

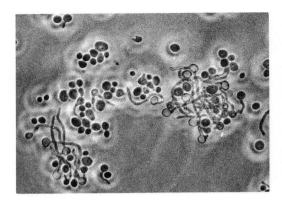

Fig. 150 Serum tube test. Cylindrical outgrowths, pseudo-germ tubes, can be seen emerging from the rounded yeast-like cells of *C. albicans*. This unstained preparation was photographed after *C. albicans* had been incubated in serum at 37°C for 1 h
    × 500

these species, *C. stellatoidea* and *C. tropicalis* are most frequently involved and the detailed identification of these and other species is usually undertaken in Reference Laboratories.

CRYPTOCOCCUS NEOFORMANS

In many countries the most common source of *Cryptococcus neoformans* is pigeon droppings, in which the yeast can survive for many weeks or months provided the excreta are shaded from ultra-violet light; pigeons themselves do not apparently suffer infection although the disease, cryptococcosis, occurs in certain wild animals and also in cats and dogs. It also occurs as one form of mastitis in cattle.

    C. neoformans is not a normal symbiont of man, and when infection takes place it is frequently subclinical and usually pulmonary, as revealed by antibody surveys. In man clinical infection declares itself as a subacute or chronic meningitis and in Britain the majority of cases of cryptococcal meningitis occur in compromised individuals particularly those with neoplasia, especially Hodgkin's disease.

C. neoformans is a yeast belonging to the same family as *Candida* and reproduces by budding, but does not produce hyphae or pseudohyphae and is characteristically surrounded by a thick capsule. Four serotypes, A–D, are recognised by agglutination testing with type specific antisera.

Only two other members of the genus, namely *C. albidus* and *C. laurentii,* have been found in human infection and these only rarely.

LABORATORY DIAGNOSIS

This depends primarily on microscopic examination of cerebrospinal fluid or other specimens to detect the characteristic spherical, capsulate yeast cells which measure from 5–20 μm in diameter.

Material should also be inoculated on to Sabouraud's medium and incubated at 37°C for at least 14 days; colonies may be apparent within a few days. Young cultures resemble those of *Staphylococcus albus,* but colonies later become tan-coloured and then brown. Colonies rapidly develop a mucoid character and India ink films of cultures show the characteristic capsulated and budding yeast cells with no evidence of a mycelium (Fig. 151).

C. neoformans is not readily differentiated from all other cryptococci on the basis of biochemical activities but it is the only species which is pathogenic for mice; within a week or two of intraperitoneal inoculation of a culture the animal dies and postmortem examination reveals gelatinous masses of *C. neoformans* in the abdomen and the brain.

Serological tests are also available, and antibody to *C. neoformans* can be detected in the majority of cases of localised skin infection, but not in cases of cryptococcal meningitis since the abundance of capsular antigen produced in the latter infection mops up the serum antibodies.

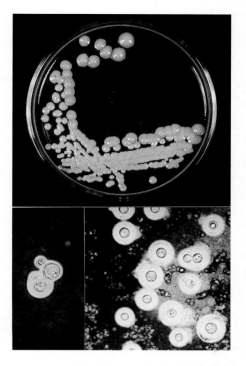

Fig. 151 *Cryptococcus neoformans.* The top photograph shows the appearance of colonies of *C. neoformans* after growth on Sabouraud's medium for 8 days at 37°C; at an earlier stage of growth the colonies resembled those of *Staph. albus* but are now heavily mucoid and would later become tan-coloured.
$\times \frac{1}{2}$
Of the two India ink films, that on the left was made from a centrifuged deposit of cerebrospinal fluid (C.S.F.) taken from a young girl suffering cryptococcosis, whilst on the right the preparation is of material taken post-mortem from the brain of a mouse 3 weeks after it had been inoculated intraperitoneally with the C.S.F.
$\times 600$
The presence of budding, thick-walled spherical cells with prominent capsules is diagnostic

However it is possible to demonstrate the presence of *C. neoformans antigen* in the CSF and serum of cases of cryptococcal meningitis using the latex particle agglutination method.

213

Further reading

As stated in the First Edition, this volume is offered as a *vade mecum* to lectures and practical classes and thus as a supplement to more complete texts; the advice which I have given to my own students throughout the years when they ask 'Which textbook should we use?' is simple: there is a plethora of texts, varying in size, on bacteriology, and students should browse through several of these to find one whose style of presentation attracts them.

Beyond the scope of textbooks on bacteriology there are several volumes which can be consulted, and these include:

Ball A P (ed) 1982 Notes on infectious diseases. Churchill Livingstone, Edinburgh

Bell D R 1981 Lecture notes on tropical medicine. Blackwell Scientific Publications, Oxford

Christie A B 1980 Infectious diseases: epidemiology and clinical practice, 3rd edn. Churchill Livingstone, Edinburgh

Garrod L P, Lambert H P, O'Grady F 1981 Antibiotic and chemotherapy, 5th edn. Churchill Livingstone, Edinburgh

Gruneberg R N 1981 Microbiology for clinicians. M.T.P. Press, Lancaster

Maurer I M 1978 Hospital hygiene, 2nd edn. Edward Arnold, London

Robertson D H H, McMillan A, Young H 1980 Clinical practice in sexually transmissible diseases. Pitman Medical Books, London

Warnock D W, Richardson M D (eds) 1982 Fungal infection in the compromised host. J. Wiley & Sons, New York

Index